NASA/SP–2013-216025

NASA's Kuiper Airborne Observatory, 1971–1995: An Operations Retrospective With a View to SOFIA

Edwin F. Erickson

Research Scientist (retired), Astrophysics Branch, MS 245-6
NASA Ames Research Center
and *Orbital Sciences Corporation*

Allan W. Meyer

Associate Scientist, SOFIA Project, MS 211-3
Universities Space Research Association
NASA Ames Research Center

National Aeronautics and
Space Administration

Ames Research Center
Moffett Field, California 94035-1000
December 2013

Cover: The Gerard P. Kuiper Airborne Observatory (KAO).

DEDICATION

This document is a tribute to the many people whose efforts produced the resounding success of NASA's airborne astronomy program—and in particular, that of the Kuiper Airborne Observatory—at Ames Research Center. In addition, we salute their family members, and our own, whose support and sacrifices were major contributions to this remarkable activity. Lastly, we recognize in particular our deceased colleagues, whose dedication to the program deserves special regard: Robert M. Cameron, Carlton M. Gillespie Jr., and James O. McClenahan.

TABLE OF CONTENTS

TABLE OF CONTENTS (cont.)

LIST OF FIGURES

Appendix C contains 106 images depicting the KAO history and participants.

LIST OF TABLES

PREFACE

In the annals of human endeavor belong accounts of unique and remarkable activities. This is such an account.

Beginning in 1965, a program of airborne astronomy observations evolved over a 30-year period at NASA Ames Research Center. Aircraft were operated for scientists who provided specialized, state-of-the-art instruments and used them to make observations not possible from the ground. Early pioneering research was done from NASA's Convair 990, U-2, and Learjet. The crown jewel of the program was the Kuiper Airborne Observatory (KAO)—a Lockheed C-141 "Starlifter" aircraft with a permanently installed 0.9-meter- (36-inch-) diameter telescope. NASA retired the KAO in 1995, after more than 21 years of effective support of the astronomical community, in order to divert its operating budget to the development of the Stratospheric Observatory for Infrared Astronomy (SOFIA).

The KAO was like no previous astronomical observatory. Its scientific achievements and excellent productivity, confirmed by independent performance assessments, rank it among NASA's highly successful astronomy programs. The remarkable record of this facility prompted us to chronicle details of its operation and to capture experiences and, in some degree, the sentiments of its participants. From these are gleaned lessons that contribute to its legacy.

We identify these lessons as *factors for success*. Principal among these are (1) co-location of science, mission, and flight-operations personnel and facilities, which enabled efficient operations and fostered close communication and teamwork; (2) proximate, minimally-restricted airspace and good local and high-altitude weather at the primary base of operations; (3) deployments for observing southern-hemisphere objects and ephemeral events; (4) a mission-systems staff sharing flight and ground-based responsibilities; (5) on-board staff and scientists capable of dealing with in-flight contingencies; and (6) operations minimizing administrative, organizational, and technical complexity. Underpinning all of these was the team attitude, a spirited emphasis of the KAO participants on achieving the basic program objective, epitomized in the oft-repeated mantra, "Let's get the data!"

KAO users were the primary advocates in achieving approval to develop SOFIA. They formed its Science Working Group, which recommended other factors for SOFIA operations that were not included in the operation of the KAO. Most significant are (7) encouragement to promptly publish observational results; (8) archiving of and access to calibrated data for the science community; (9) attractive procedures for general investigators to obtain data with any existing science instrument, and (10) scientific direction of the observatory. Explanation of, and rationale for, these ten factors, along with numerous others, are provided.

We hope this historical review of KAO operations will reignite pleasant memories for its alumni and others familiar with the program. To the many current and future SOFIA participants, we commend consideration of the KAO experience presented here. For all readers, we hope this retrospective will be enlightening and enjoyable.

Edwin F. Erickson   Allan W. Meyer   December 2013

ACKNOWLEDGMENTS

We are grateful for thc comments on and contributions to this document by some of our many friends from the KAO era: Dana Backman, Larry Caroff, Jim Cockrell, Chuck Connors, Jeff Cuzzi, Edna DeVore, Wendy Dolci, Ted Dunham, Paul Fusco, Bob Gehrz, John Graybeal, Mike Haas, Warren Hall, Martin Harwit, Randy Hobbs, Hans Kaercher, Nans Kunz, Dan Lester, Tom Roellig, and Xander Tielens who suggested this review. We also thank Lisa Chu-Thielbar, Kevin Martin, and Leon Shen for their insights on particular issues, and Lynn Albaugh for retrieving many of the images from the Ames photo archive. Pam Marcum, Erick Young, and particularly Dan Lester and Nans Kunz provided valuable comments on drafts of the manuscript.

Excellent editing and formatting were done by Cathy Dow. At Ames, Lori Munar and Nikki Isaac assisted with details required for publication, and Melody Miles, John Woebcke and Raul Dingle handled printing. Ed Erickson appreciates especially Tim Lee and Jessie Dotson for their consistent encouragement.

This work was partially supported by the SOFIA and NESC programs of the National Aeronautics and Space Administration.

ACRONYMS

AAS	American Astronomical Society
ADAMS	Airborne Data and Management System
ALMA	Atacama Large Millimeter Array
ARC	Ames Research Center
ASD2VME	Avionics Serial Data to VersaModular Eurocard interface
ASP	Astronomical Society of the Pacific
ATRAN	Atmospheric Transmission calculation program
AU	Australia
BMAC	Boeing Military Airplane Co.
BMFT	Bundesministerium für Forschung und Technologie
CADC	Central Air Data Computer
CanariCam	Facility mid-infrared imaging spectrograph
CARA	Center for Astrophysical Research in Antarctica
CASIMIR	CAltech Submillimeter Interstellar Medium Investigation Receiver
CCD	Charge-Coupled Device, in integrated detector arrays
CEA	Atomic Energy Commission (FR)
CGS	Cryogenic Grating Spectrometer
CII	Singly ionized carbon
CMBR	Cosmic Microwave Background Radiation
COBE	Cosmic Background Explorer satellite
CPI	Consumer Price Index
CS	Civil Service
CSO	Caltech Submillimeter Observatory
DARA	Deutsche Agentur für Raumfahrtangelegenheiten
DE	Germany
DFRC	Dryden Flight Research Center
DFVLR	Deutsche Forschungs- und Versuchsanstalt für Luft- und Raumfahrt
DLR	German Aerospace Center
E&PO	Education and Public Outreach
ESA	European Space Agency
ESO	European Southern Observatory
FAA	Federal Aviation Administration
FIFI-LS	Field Imaging Far-Infrared Line Spectrometer
FL	Flight Level (altitude) in hundreds of feet
FOC	Full Operational Capability
FORCAST	Faint Object InfraRed CAmera for the SOFIA Telescope
FOSTER	Flight Opportunities for Science Teacher Enrichment
FR	France
FSO	Flight Safety Office
FTE	Full-Time Equivalent
GI	Guest Investigator: a scientist observing with but not part of an instrument team
GPO	Government Printing Office
GREAT	German Receiver for Astronomy at Terahertz Frequencies
GSFC	Goddard Space Flight Center
HAWC	High-resolution Airborne Wideband Camera

HI-FI	Heterodyne Instrument for the Far Infrared
HIPO	High-speed Imaging Photometer for Occultations
HST	Hubble Space Telescope
IG	Inspector General
INS	Inertial Navigation System
IRAC	Infrared Array Camera
IRAS	Infrared Astronomy Satellite
IRS	Infrared Spectrometer
IRTF	Infrared Telescope Facility
IRTS	Infrared Telescope in Space (JP)
ISM	Interstellar Medium
ISO	Infrared Space Observatory (ESA)
JP	Japan
JPL	Jet Propulsion Laboratory
JWST	James Webb Space Telescope
KAO	Kuiper Airborne Observatory
Kepler	Satellite to Search for Earthlike Planets
KOSMA	Kölner Observatorium fűr Submm Astronomie
LAS	Lockheed Aircraft Systems
LAT	Large Airborne Telescope
LDR	Large Deployable Reflector
LIGO	Laser Interferometer Gravitational Wave Observatory
MANGHH	Maschinenfabrik Augsburg-Nürnberg
MD	Mission Director
MIRS	Mid Infrared Spectrometer
MIT	Massachusetts Institute of Technology
MPE	Max Planck Institute for Extraterrestrial Physics
MPIA	Max Planck Institute for Astronomy
NASA	National Aeronautics and Space Administration
NASTRAN	NASA finite element structural analysis software program
NESC	NASA Engineering and Safety Center
NICMOS	Near Infrared Camera and Multi Object Spectrometer
NL	Netherlands
NOAA	National Oceanographic and Atmospheric Administration
OMB	Office of Management and Budget
OSM	Oscillating Secondary Mirror
PACS	Photodetector Array Camera and Spectrometer
PAH	Polycyclic Aromatic Hydrocarbon
PCLS	Portable Chopped Light Source
PEAKR	Software enabling automated boresighting of science instruments to telescope camera
PFC	Passive Flow Control
PI	Principal Investigator
QA	Quality Assurance
PYTHON	CMBR Submillimeter Polarimeter
RIT	Rochester Institute of Technology

RTCA	Radio Technical Commission for Aeronautics
SAFARI	SpicA FAR-infrared Instrument
SAFIR	Single Aperture Far Infrared Observatory
SAO	Smithsonian Astrophysical Observatory
SFH	Successful Flight Hours
SHARC	Submillimeter High-Angular Resolution Camera
SHARP	SHARC CII Polarimeter
SI	Science Instrument
SIRTF	Space Infrared Telescope Facility
SIS	Superconductor-Insulator-Superconductor
SOFIA	Stratospheric Observatory for Infrared Astronomy
SPARO	Submillimeter Polarimeter for Antarctic Remote Observing
SPICA	Space Infrared Telescope for Cosmology and Astrophysics (JP)
SR & QA	Safety, Reliability, and Quality Assurance
SPIRE	Spectral and Photometric Imaging Receiver
Spitzer	Space Infrared Telescope (formerly SIRTF)
SRON	Netherlands Institute for Space Research
SSOP	SOFIA Science Operations Program
SSWG	SOFIA Science Working Group
STScI	Space Telescope Science Institute
SUA	Special Use Airspace
SWAS	Submillimeter Wave Astronomy Satellite
SWS	Short Wavelength Spectrometer
TDADS	Telescope Data Acquisition and Display System;
TIPS	Telescope Inertial Pointing System
T-ReCS	Facility mid-infrared imaging spectrograph
TSC	Telescope System Controller
UNSW	University of New South Wales
USAF	United States Air Force
USRA	Universities Space Research Association
WINDO	Software for examining availability of astronomical objects for flights
WISE	Wide-Field Infrared Survey Explorer (satellite)
WMAP	Wilkinson Microwave Anisotropy Probe (satellite)

1. INTRODUCTION

Astronomy from aircraft allows observations from the near ultraviolet to millimeter wavelengths, an exceptionally broad range, much of which is inaccessible from ground-based sites. Likewise, the mobility of an airborne observatory enables access to the entire celestial sphere, and to cloud-free observing of short-lived, highly localized astronomical events. Neither ground- nor space-based telescopes provide this wavelength coverage and/or mobility.

As with ground-bound observatories, an airborne observatory permits routine opportunities for upgrading science and mission systems as technologies advance and observing opportunities arise, and for use of a wide variety of science instruments, hands-on training of young scientists, and participatory involvement of educators and media people. Support facilities and personnel can be conveniently located where an airborne observatory is based—a powerful factor for efficient operation that is denied to many ground-based and all space-based observatories. Finally, an airborne telescope permits convenient public visiting opportunities wherever the aircraft lands.

Realizing the unique potential of an airborne observatory requires unusual operational procedures. Effective ones were developed in a program of airborne infrared astronomy at NASA Ames Research Center. To begin, a brief description of the remarkable pioneering efforts that inspired the concept of a dedicated facility, made manifest in the Kuiper Airborne Observatory (KAO), is provided (refs. 1-3). After a description of the relevant effects of the Earth's atmosphere, there follows a sketch of the KAO development and a summary of some of its science highlights. The subsequent extended chronicle of the unique KAO operation includes achievements and metrics demonstrating that the KAO was successful indeed. "Success" in NASA parlance means safely achieving a level of science productivity that approaches the maximum possible for available resources. This record of KAO operations leads naturally to identification of numerous *factors for success* that characterized the program.

Following this account is a brief description of SOFIA (refs. 4, 5), including rationale, history, and expectations and recommendations of its planners, most of whom were KAO users. They incorporated explicitly or implicitly assumed many of the KAO *factors for success* in planning SOFIA, in recognition that the unique KAO experience was the most comprehensive model for effective SOFIA operations.

Preceding a short summary, some poetic vignettes reveal the participants' fondness for their cherished KAO.

Two appendices supply detailed information summarized in the body of the text. A third appendix consists of images depicting many of the KAO participants and program highlights, tracing the evolution of the activity chronologically throughout its 21-year operational lifetime. This compendium of photos exhibits the real-life character of this exceptional, exciting NASA astronomy program.

2. FORERUNNERS OF THE KAO

The history of astronomy from airplanes from its beginnings in the 1920s has been well described by Dolci (ref. 6). From the mid-1960s at NASA Ames Research Center, ad-hoc astronomical observations from aircraft were carried out as part of a broad-based "Airborne Science" program, headed by Dr. Michel Bader, chief of the Space Science Division. Robert Cameron, a branch chief in the Flight Operations Directorate, was responsible for coordinating much of the astronomy activity.

2.1 THE CONVAIR 990

In 1966–1967, airborne measurements were extended beyond the typical visible-light observations of comets and solar eclipses by Gerard P. Kuiper and Frederic F. Forbes, who used the Ames Convair 990 "Galileo" (fig. 1) to measure the near-infrared spectra of Venus (ref. 7).

Until then the clouds of Venus had been widely assumed to contain water, but confirming observations from the ground were precluded by water vapor in the Earth's atmosphere. By flying above most of this moisture, Kuiper and Forbes obtained data that showed the clouds of Venus were dry! The observations also demonstrated the potential for using sophisticated infrared instruments on aircraft to obtain measurements not possible from ground-based sites.

Figure 1. The Convair 990 "Galileo" circa 1967. This aircraft was used for a variety of research.

2.2 LEARJET OBSERVATORY

In the late 1960s, Frank J. Low initiated far-infrared observations from aircraft using bolometer detectors he had developed and the NASA Ames Learjet (fig. 2). The clever 12-inch open-port telescope developed by Low and Carl Gillespie included the first chopping secondary mirror to suppress noise from fluctuating power ("sky noise") emitted by the atmosphere, and to reduce noise caused by extraneous radiation from the telescope (ref. 8).

These developments presaged the possibility of routine observations in the then largely unexplored broad spectral range encompassing the near-, mid-, and far-infrared and submillimeter wavelengths. This realization, and the previous successful observations of comets and eclipses from airplanes at mostly visible wavelengths, made a strong case for an enhanced capability. Recognizing this, and based on their experience in the development and operation of platform aircraft for scientific research, Ames management successfully proposed a new facility with a larger telescope, and work began in mid-1969 on what would become the Kuiper Airborne Observatory.

As this development proceeded, further measurements from, and capabilities on, Ames aircraft continued to demonstrate the potential of this discipline. In the early 1970s, Ames scientists led development of an improved Learjet telescope (ref. 9) to accommodate visiting and in-house instrument teams. Although the Lear was used for a variety of non-astronomy research, it was called the Learjet Observatory, and was used by several groups to observe solar system, stellar, and Interstellar Medium (ISM) objects until about 1976 and (with decreasing frequency) beyond that time, with the last attempted observations in 1997. Some of the important results were luminosities of star-forming molecular clouds, evidence for concentrated sulfuric acid droplets as the major constituent of the Venus clouds, and first observation of the important ISM-cooling C^+ line at 158 microns (μm) wavelength.

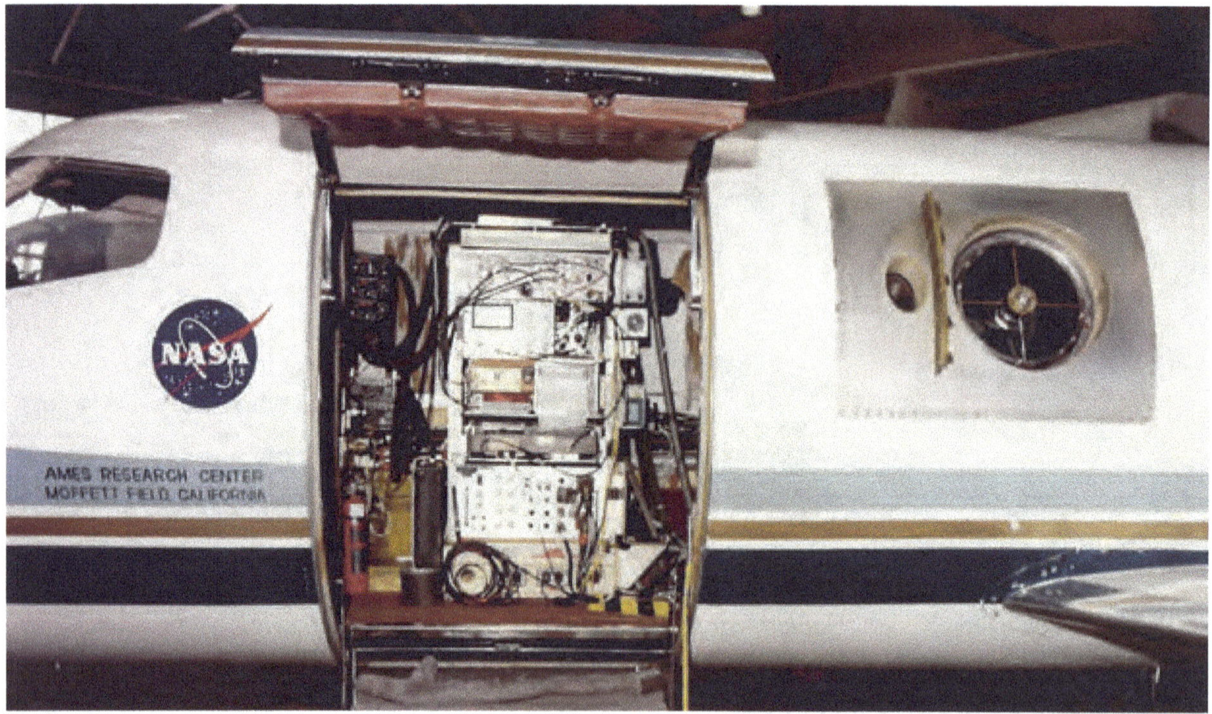

Figure 2. The Learjet Observatory. The 12-inch telescope was removable, permitting use of the plane for other research programs.

2.3 THE U-2 AIRCRAFT

In the early 1970s, a group from the University of California at Berkeley, led by Richard Muller and George F. Smoot, discovered the dipole distribution of the Cosmic Microwave Background Radiation using NASA Ames high-altitude U-2 aircraft (fig. 3), which was usually used for atmospheric research. The results were interpreted as gravitationally induced motion in the local group of galaxies, implying that the large-scale cosmic distribution of matter is dominated by vast voids punctuated by super-clusters of galaxies. This distribution is now understood as a relic of density inhomogeneities generated in the early universe, as shown by stunning data from follow-on facilities (e.g., NASA's Cosmic Background Explorer (COBE), and the Wilkinson Microwave Anisotropy Probe (WMAP)) initially inspired by the U-2 results.

Successful use of the U-2 by the Berkeley group was an additional demonstration of the effective infrastructure at NASA Ames for enabling exciting research in airborne astronomy.

Figure 3. The U-2 aircraft. This plane was used for a variety of high-altitude research.

3. THE ENABLING ATMOSPHERE

Qualitatively, the Earth's atmosphere can lower the sensitivity of astronomical observations by blurring the image (seeing), by attenuating the signal, and by generating the noise seen by the detectors in the science instrument. The three most significant atmospheric constituents affecting infrared observations are the triatomic molecules of (a) ozone, which is high in the atmosphere and least significant; (b) carbon dioxide, which is uniformly mixed and important out to wavelengths of roughly 16 µm; and (c) water, which is highly stratified and is the dominant malefactor over most of the infrared spectrum.

The stratification of water vapor overhead is described by the overburden or column-depth ***W*** in µm, as a function of altitude. Figure 4 shows measurements of ***W*** and temperature for a northern mid-latitude in winter (refs. 10, 11). With increasing altitude, the temperature drops until the tropopause is reached, remains constant, and then rises in the stratosphere. The tropopause altitude varies with time of year and location; it is typically lowest in the winter and becomes lower approaching the poles. As seen in this example, and generally, the overburden ***W*** decreases rapidly with increasing altitude as the tropopause is entered and declines more slowly after that. For the data plotted in figure 4, at an altitude of 40,000 feet, **W** is approximately 8 µm. At good ground-based sites, the value is, at best, about 25 times higher.

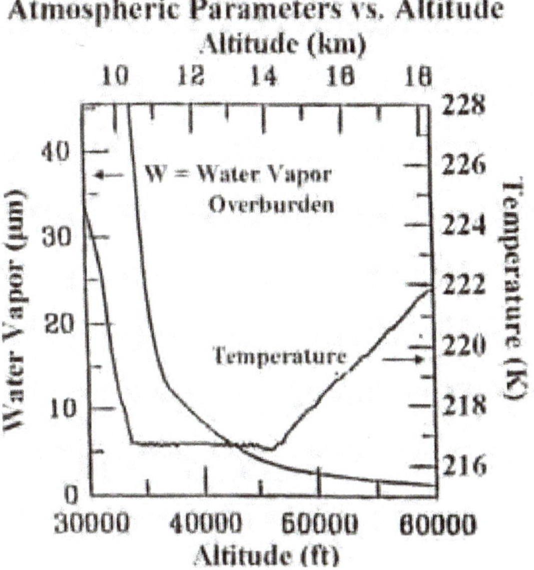

Figure 4. Temperature and overhead moisture versus altitude.

Structure of the water molecule makes the absorption very wavelength dependent. At any wavelength, the absorption varies exponentially with **-W**. Transmission at low spectral resolution, including H_2O, CO_2, O_3, and aerosols (important at visible wavelengths) is shown in figure 5. The ATRAN program (ref. 12), written by Steve Lord while a postdoctoral associate at Ames, calculates the transmission using atmospheric and science-instrument parameters for a given observation. Effects of water vapor on airborne infrared observations are described by Erickson (ref. 13). It is fortunate that the atmosphere enables airborne infrared observations.

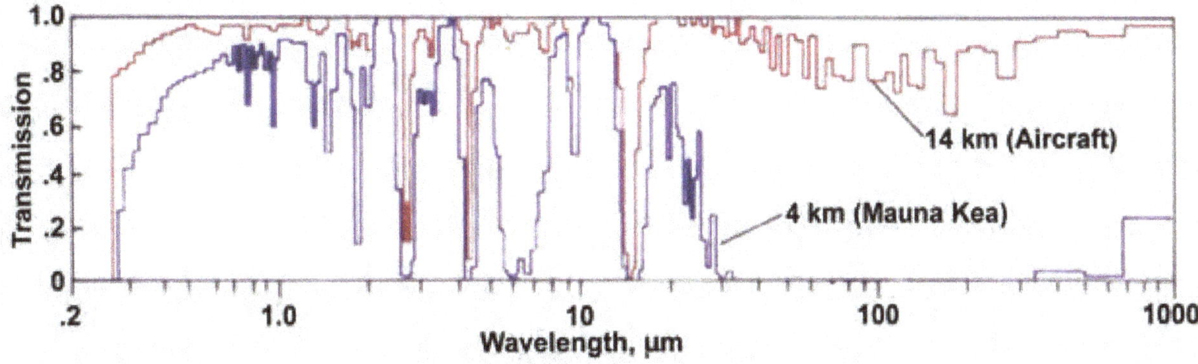

Figure 5. Wavelength dependence of representative atmospheric transmissions at two altitudes.

4. THE KUIPER AIRBORNE OBSERVATORY

4.1 DEVELOPMENT

Based on the potential demonstrated by the earlier airborne infrared astronomy accomplishments and the improved understanding of atmospheric effects described previously, feasibility studies for a dedicated airborne observatory with a 36-inch telescope were initiated and refined at Ames in the late 1960s. The concept was successfully promoted at NASA by Ames Director Dr. Hans Mark, Michel Bader, and Robert Cameron. Specifications were drafted in 1967 and finalized in 1970. The system design and plans for operation are described in a paper by Cameron, Bader, and Robert Mobley (ref. 14); Mobley was a senior engineer at Ames.

From competitive bids, the Ames project team selected Fecker Systems of Owens, Illinois, to develop the facility in a partnership with Lockheed Aircraft Systems (LAS) in Ontario, California. Fecker built the telescope according to the unusual design they had proposed. LAS modified the Lockheed C-141A "Star-Lifter" 4-engine jet cargo plane—a demonstration model that NASA had purchased for a bargain price from the manufacturer (fig. C3). In 1972, due to contractual difficulties, the initial assembly and testing of the telescope was moved to Ames (fig. 6), where it was completed by a team comprised of civil service (CS) and contractor personnel.

Figure 6. The KAO telescope during initial assembly and testing at NASA Ames.

The development is outlined in figure 7. It is remarkable that the aircraft was modified and the telescope installed in about 8 months, and structural/aerodynamic flight tests were completed with six flights in 1 month. The full cost, including CS labor, was about $110M (in 2013 dollars) (see section 4.10).

1969	1970	1971	1972	1973	1974
A M J	M J J A S O N D	J F M A M J J A S O N D	J F M A M J J A S O N D	J F M A M J J A S O N D	J F M A M J J

Fecker Designs & Builds Telescope
NASA Procures C-141 from Lockheed
Telescope assmbled & ground tested at Ames
Aircraft, Telescope delivered to LAS
Aircraft modified & Telescope installed
Structural Flight Tests at DFRC
Door drive & wing fairing modifications at Ames
Functional flight testing & debugging at Ames
Science Flights Commence

Figure 7. Development timeline of the Kuiper Airborne Observatory.

4.2 ASTRONOMY PROGRAM SYNOPSIS

Kuiper's outstanding work using the Ames Convair 990, his conclusions regarding the use of aircraft for infrared observations, and his encouragement in developing the new facility were honored by dedicating it in his name. His wife christened it the Gerard P. Kuiper Airborne Observatory (KAO) soon after his death in December 1973. The observatory is shown on the cover of this document and in figures 8 and 41.

The KAO was based at, and operated from, Ames in support of research programs from mid-1974 until the fall of 1995. Over its 21-year operational lifetime, KAO research involved more than 600 investigators and produced over 50 Ph.D. theses and 1,000 scientific and technical papers. Observing time and science instruments were selected by annual peer review. Two airborne astronomy symposia (refs. 1,3) were held at Ames, celebrating the 10th and 20th years of the KAO operation. Proceedings of these events document many of the outstanding results of the airborne astronomy program, including observations from visible to millimeter wavelengths. An excellent summary of the science program, productivity, and science-community participation through 1990 was published by Larson (ref. 2).

Figure 8. The KAO poised for an observing mission in front of main hanger N-211 at NASA Ames.

4.3 SCIENCE HIGHLIGHTS

Many remarkable scientific findings are described in the roughly 1,000 publications resulting from KAO observations. These were obtained on both solar system objects and on the more distant objects generally considered to be the realm of astrophysics. Note that the KAO program was funded entirely by the Astrophysics Division at NASA Headquarters. The four major publication categories were galactic astronomy (57 percent), planetary science (25 percent), extragalactic astronomy (9 percent), and instrumentation (9 percent) (ref. 2). Papers describing some of the results can be found in references 1 and 3. A few highlights are listed below.

Solar system: discovery of the rings of Uranus; measurement of the intrinsic luminosities of Jupiter, Saturn, and Neptune; discovery of water in Jupiter's atmosphere and comets; discovery of Pluto's atmosphere; evidence for a distribution of small ice particles on Saturn's rings.

Astrophysics: explorations of a major new component—photodissociation regions (neutral atomic gas excited by non-ionizing stellar radiation)—in the Interstellar Medium (ISM); discovery of over 70 spectral features arising from atoms, ions, molecules, and grains in the ISM; discovery of star-forming cores in isolated dark clouds called Bok globules; discovery of far-infrared luminosities of normal galaxies comparable to their visible luminosities; early evidence for hot stars and a black hole in the Galactic Center of the Milky Way; measurement of iron, cobalt, nickel, and argon manufactured in supernova SN1987A; first detection of astronomical far-infrared lasers; and discovery and identification of mid-infrared spectral features from prebiotic polycyclic aromatic hydrocarbon (PAH) molecules.

Many of these and other important KAO results were completely unanticipated. In addition to astronomical observations, the KAO occasionally supported other research activities, some in a dedicated mode using the telescope and some in piggyback mode for atmospheric research. Routine radiometric monitoring of atmospheric water vapor for the observatory resulted in a new technique for warning of imminent clear-air turbulence.

4.4 OPERATIONS

4.4.1 Flight Program

Records of KAO flight operations from FY1972 through FY1996 are given in table A1 of Appendix A. Table 1 summarizes these results for the period from FY1974 to FY1995 when the facility was flying astronomy missions. A typical astronomy flight provided 6¼ to 6½ hours of observing, from opening the cavity door at 35,000 feet until beginning descent. The scheduled "block-to-block" duration was normally 7.5 hours, limited by NASA pilot restrictions, and consistent with the fuel-load limitation on hours at altitude. Flight hours shown in table 1 correspond to these block-to-block durations of the aircraft operation. The small fraction of non-astronomy research (consisting of atmospheric studies, NASA, and military programs) done with the KAO is noted in the table as "Other Research." Excluding the ramp-up year 1974, the average annual number of hours flown by the plane was 635.

Table 1. KAO Flight Operations, FY1974–FY1995

	Flights		Flight Hours	
Flight Category	#	%	#	%
Aircraft Maintenance	259	11	332	3
Pilot Proficiency	144	6	439	3
Engineering	121	5	533	4
Ferry	135	6	716	6
Sum, Non-research	**659**	**29**	**2020**	**16**
Aborted (Research)	136	6	392	3
Other Research	39	2	222	2
Astronomy Research	1,424	63	10240	80
Sum, Research	**1599**	**71**	**10854**	**84**
Totals	**2258**	**100**	**12873**	**100**

Figure 9, based on data from table A1, plots the annual numbers of astronomy flights, astronomy flight hours, and aborted flights from all—mostly research—categories. Large year-to-year variations were due primarily to major upgrade and maintenance activities. The rise in the rate of aborted flights (1981–1983) may have been due to the increasingly sophisticated technologies being implemented in the mission systems and in the (all) user-supplied science instruments. For example, at the beginning of the program, computer codes for the KAO tracking system consisted of a few hundred instructions, loaded from a punched paper tape. Data were commonly plotted on line printers. As computer speeds and memory capacities increased, larger instruction sets with more functionality were employed, with concomitant possibilities for errors. Beginning in the late 1970s, science-instrument detector systems began expanding from single to multiple discrete elements, and eventually (in some cases) to integrated semiconductor arrays. Accompanying the increased capability was somewhat increased susceptibility to malfunction.

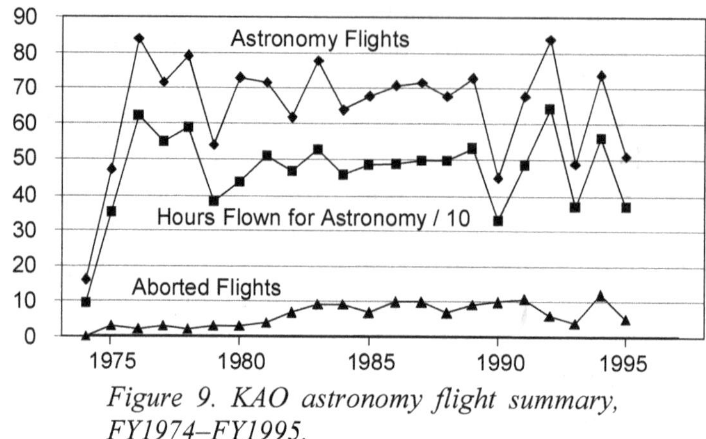

Figure 9. KAO astronomy flight summary, FY1974–FY1995.

Prompt deployment of technical improvements was always encouraged as part of the program's goals. Of course such improvements dramatically increased data acquisition rates and observing productivity per successful flight hour throughout the KAO's lifetime.

4.4.2 Flight-Efficiency / Reliability Statistics

Typically 80 research flights were planned at the beginning of each year when no major aircraft maintenance or system upgrades were scheduled. From statistics for flight operations, Appendix A estimates that the fraction of research flights flown / planned was 0.88. On average, about 76 research flights were flown with 70 (or 92 percent) being completed each year (FY1975–FY1995). Details of efficiencies and lost flight time are discussed below.

Roughly 87 percent of flight time on astronomy missions (approximately 6½ hours out of 7½ hours) was dedicated to astronomy, with the aircraft at altitude and the telescope cavity door open. The number of hours flown for astronomy averaged 488 per year, so the annual number of observing hours averaged about 424 (488 × 0.87).

Data acquisition efficiency for astronomy observations at operating altitudes typically ranged from 70–85 percent, depending on maturity of the science instrument and team, flight-plan constraints, time spent in turns between flight legs, and time spent acquiring sources and setting up telescope and instrument parameters. Naturally observations of short-duration events requiring specific locations of the aircraft (e.g., occultations and eclipses) usually had much lower observing duty cycles. The efficiency numbers are summarized in table 2.

Three categories of inefficiencies were recorded: flights delayed by one or more days, flights aborted after takeoff, and flights cancelled before takeoff. Delayed flights were flown at a later date, typically during the same flight series, independent of cause. Statistics from FY1981 through FY1995 for these categories are shown in table 3 with causes identified; consistent earlier data could not be located. The miscellaneous category includes weather, personnel problems, unexpected airspace restrictions, etc. Eleven (over half) of the cancellations due to

Table 2. Typical Efficiencies for KAO Observations	
Research Flights Flown / Research Flights Planned	0.88
Research Flights Completed / Research Flights Flown	>0.92
Research Flight Hours / Total Flight Hours in All Categories	>0.82
Fraction of Block-to-Block Flight Time for Astronomy	~0.87
Observing Efficiency from End of Climb-Out to Start of Descent	0.70–0.85

Table 3. Causes of Research Flight Interruption, FY1981–FY1995

Flight	Instrument	Aircraft	Telescope	Misc.	Totals
Delayed	27	63	43	13	146
Aborted	22	28	75	3	128
Cancelled	28	24	17	17	86
Totals	77	115	134	33	359
%	21	32	37	9	100

"Experiment" occurred in 1987, possibly because several science instruments were rapidly reconfigured to observe Supernova 1987A that year. Generally the science instruments were the most reliable system element, followed by the aircraft, and then by the telescope. Some of the least reliable telescope systems were the air-bearing compressors, the cavity environmental control system, the secondary mirror chopping mechanism, and the telescope servo/drive electronics.

Aborted flights were those terminated early due to problems that precluded data acquisition. The numbers of aborted flights listed in table 3 include those events for all the flight categories listed in table 1. Of course, most aborts occurred on research flights because they were far more numerous than those in other categories, and because of the need for the sophisticated science research and science mission systems (telescope, etc.) to perform adequately. Usually on aborted flights, some time was spent diagnosing the problem, modifying the flight plan, and flying back to the operations base from an unanticipated turnaround point. Generally, research flights were not aborted if acquisition of useful data could continue despite malfunctioning equipment or other problems. Often a problem occurring in flight was solved with a relatively small loss of observing time, thereby avoiding an abort; this fact is a tribute to the knowledgeable personnel operating the science mission systems and instruments. Malfunctions sometimes occurred after some hours of observing, so that useful data were often acquired on flights recorded as aborts. Hence the significance of flight hours listed for aborted flights in table 1 is obscure.

The uniqueness and complexity of the KAO operation makes it difficult to identify analogous programs whose operational efficiencies can be reasonably compared with those cited in table 2. Airborne earth-science programs, for example, have much lower flight frequency and takeoff-time criticality, and have no dedicated mission system as complex as the telescope. It is certain, however, that these excellent numbers for the KAO, approximately 85-percent efficiencies, resulted from a variety of contributing factors alluded to above and discussed next.

The flight rate was limited by available funding and by the CS workforce allocated to the program. Proposed increases were never successful, in significant part because of the increase in cockpit-crew staffing that was deemed necessary by the Flight Operations Directorate (see sections 4.7 and 4.9).

4.4.3 Astronomy Missions

Science instruments were typically installed for a 1- or 2-week flight series. Usually 40 weeks of science operations were scheduled annually, including multiple deployments to remote sites. Typically two flights were scheduled per week, but rates of three flights per week were often sustained while on deployment. A few times, four flights were flown in a week. As seen in table 4, during the last 10 years of operation about 40 percent of astronomy research flights were made while the aircraft was deployed to remote sites, whereas about 18 percent were remotely staged during the previous decade. Overall, 73 percent of the astronomy flights were flown from Ames.

Table 4. KAO Astronomy Year-by-Year Flight Summary

Fiscal Year	Flights	Ames	Hawaii	New Zealand	Australia	Other	Location of Other	% not from Ames
1974	16	16						0
1975	47	27	20					43
1976	84	84						0
1977	72	61			11			15
1978	79	69	8			2	Samoa	13
1979	54	38	13			3	Samoa	30
1980	73	63	8			2	Panama	14
1981	72	61	7			4	Japan	15
1982	62	49	13					21
1983	78	58	3		15	2	Guam	26
1984	64	54	10					16
1985	68	53	15					22
1986	71	41	5	25				42
1987	72	58	6	8				19
1988	68	41	5	22				40
1989	73	42	5	26				42
1990	45	25	3	17				44
1991	68	42	10	16				38
1992	84	63	5	16				25
1993	49	16	10	21		2	Ecuador, Chile	67
1994	74	45	10	11	6	2	Brazil	39
1995	51	23	20	8				55
Totals	1424	1029	176	170	32	17		28
%	100	72	12	12	2	1		28

4.4.4 Deployments

Operations from remote locations were a major factor in the success of the KAO. As shown in table 4, these operations comprised over a quarter of all astronomy missions. They consisted of (1) relatively routine observations of objects in the southern sky (not observable from Ames because the elevation range of the telescope was 35–70 degrees); these involved extended deployments and multiple instrument teams, and (2) observations of ephemeral events—usually, but not always, with a single instrument. The following text describes two early examples that typify the two types of excursions.

One of the longest deployments—20 flights—was to Hawaii in 1975, the year after the KAO began flying astronomy missions. Several flight series were flown sequentially for different instrument teams. The mission was based at Hickam Air Force Base where good aircraft support and maintenance (from the Air Force) was available. The aircraft was parked on the tarmac, so line ops were sometimes compromised, for example by overheating of some mission systems (e.g., the telescope gyros), or were precluded by inclement weather. A couple of large trailers on the tarmac located near the aircraft were available for ground crew and instrument team activities. Participants stayed in Honolulu hotels some miles from Hickam, a somewhat awkward arrangement. Heavily trafficked commercial flight lanes restricted flight planning somewhat. Nevertheless, the expedition was highly successful, resulting in a number of unique scientific findings.

In March of 1977, the KAO flew to Perth, Australia. It was based there to observe a star that would be occulted by Uranus, an event expected to last minutes and to be observable from a location over the Indian Ocean. The C-141's inertial navigation system (accurate to within a nautical mile) proved the prediction correct, so the event was successfully observed. More than obtaining the intended information on Uranus' atmosphere, the measurements revealed a previously undiscovered system of rings circling the planet!

Beginning in 1986, Christchurch, New Zealand, became the favorite site for extended deployments to the southern hemisphere. From there, in addition to availability of the majority of the sources in the Galactic plane (including the Galactic Center) not observable from Ames, the occurrence of Supernova 1987A provided an exceptionally interesting target for several years as its envelope expanded and dimmed. At the 43.5-degrees south latitude of Christchurch, the local winter nights are long and the tropopause (see section 3) is low. There are no restricted "no-fly" zones off the islands and little commercial air traffic to complicate flight planning. Weather on the ground is moderate. The hotel for participants was in easy walking distance of the airport where the plane was based, as were the staging facilities for the U.S. Antarctic Program, where support for science activities and a good cafeteria (!) were available. Facilities at the airfield next to the plane consisted of two trailers for the observatory personnel, and one for each of two instrument teams (one flying, one arriving). Overall the onsite observatory staff (including pilots, ground crew, telescope crew, navigator, and managers) numbered about 20. Ames provided relief for them at least once during a typically 6- to 8-week deployment. The telescope crew was pretty lean because the technicians who could service the telescope and the observatory managers present also flew in operational roles.

The most intense KAO deployment occurred in 1994 to Melbourne Australia, when six flights in about a week were taken, accommodating multiple instruments to observe impacts of comet Shumaker-Levy on Jupiter. Of course for this expedition a much larger observatory crew—about 40 people—was required.

4.5 EFFICIENCY-ENHANCING FEATURES

4.5.1 Flight Planning

Efficient and flexible flight planning was critical to program effectiveness. Most observing flight plans were produced by investigator teams collaborating with staff navigators, using two software tools provided by the observatory: (1) WINDO (ref. 15), which calculates times when objects are in the 35–75° elevation range of the telescope at a given time of year and location of the plane, and (2) KNAV (ref. 16), which computes flight legs (aircraft heading and position, and object elevation versus time) and plots the predicted ground track for a given sequence of astronomical objects.

These programs allowed generation of efficient flight plans for a dozen or more objects on a flight, although fewer objects were observed on most flights. These software tools were also available and used in flight when unexpected circumstances required real-time modification of the original flight plan.

Versions of both WINDOW (later versions were called WINDO) and KNAV were upgraded over the life of the KAO program, and were made available to investigators for use on their own computers, which substantially facilitated the flight planning process. Investigators typically iterated their plans using KNAV, adjusting priorities, flight-leg durations, and object sequences to optimize the result to their satisfaction.

Figure 10. Western U.S. Special Use Airspace (SUA) areas. Arrows show locations of Ames (A) and Palmdale (P).

Typically a day or so before a flight, the investigator teams would provide their KNAV flight plan to the Ames navigators. They would refine and finalize the plan using current wind predictions and airspace restrictions (fig. 10) and submit the plan to the local air traffic control office. On average, the total effort needed to prepare the flight plan was 8–12 person-hours.

KNAV included world-wide geography and incorporated Special Use Airspace (SUA) areas (no-fly zones), shown in figure 10 for the western United States. The density of SUA areas decreases north of Ames and south of Ames. Permission to overfly SUA areas is controlled by individual agencies, so a separate permission is required for each. Most, but not all, SUA areas may be violated some of the time, but they may be closed with little warning. Experienced KAO flight planners and navigators shunned all but the most commonly available SUA areas, as well as the commercial flight lanes.

Ascent and descent legs were chosen to position the plane to optimize flight plans. Those originating from Ames typically started observations centered about 150 miles north of Moffett Field. This strategy allowed minimal incursions into SUA areas, thereby minimizing the loss of observing time due to deviations from the filed flight plans. Figure 11 is a KAO flight plan from 1995, reconstructed using the current equivalent of KNAV. It demonstrates the careful avoidance of SUA areas while flying over the western U.S.

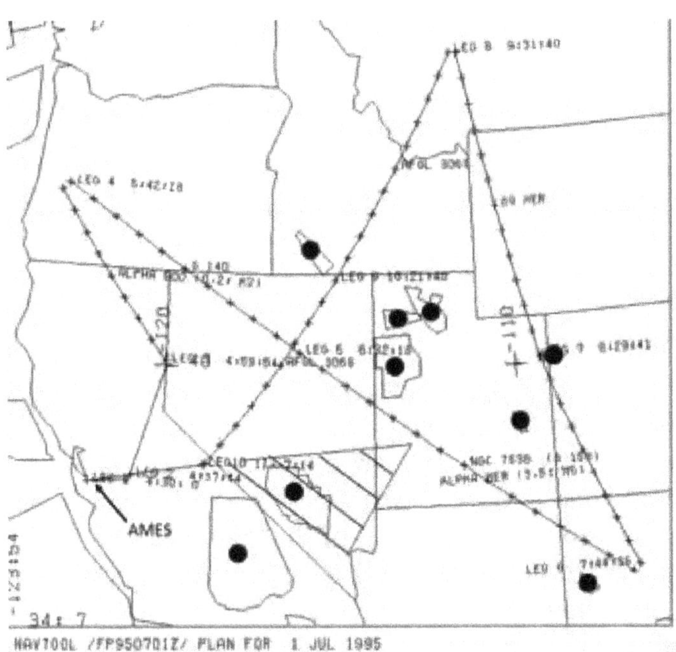

Figure 11. A KAO flight plan. Originating at Ames, this flight avoided the nearby SUA areas that are marked with black dots. The cross-hatched triangle near bottom center encloses an additional zone that the Ames navigators excluded from flight plans.

Flights over Canada and Mexico required special clearances and were generally avoided. Flying over Mexico was often not desirable because of the frequently higher water vapor at mid-latitudes, as described by Haas and Pfister (ref. 17). Flying over Canada was normally not attempted anyway because the distance to the border typically precluded efficient flight planning.

A superposition of 48 KAO flight plans based at Ames is shown in figure 12. This figure demonstrates the general patterns, and avoidance of SUA areas, international borders, and commercial flight lanes.

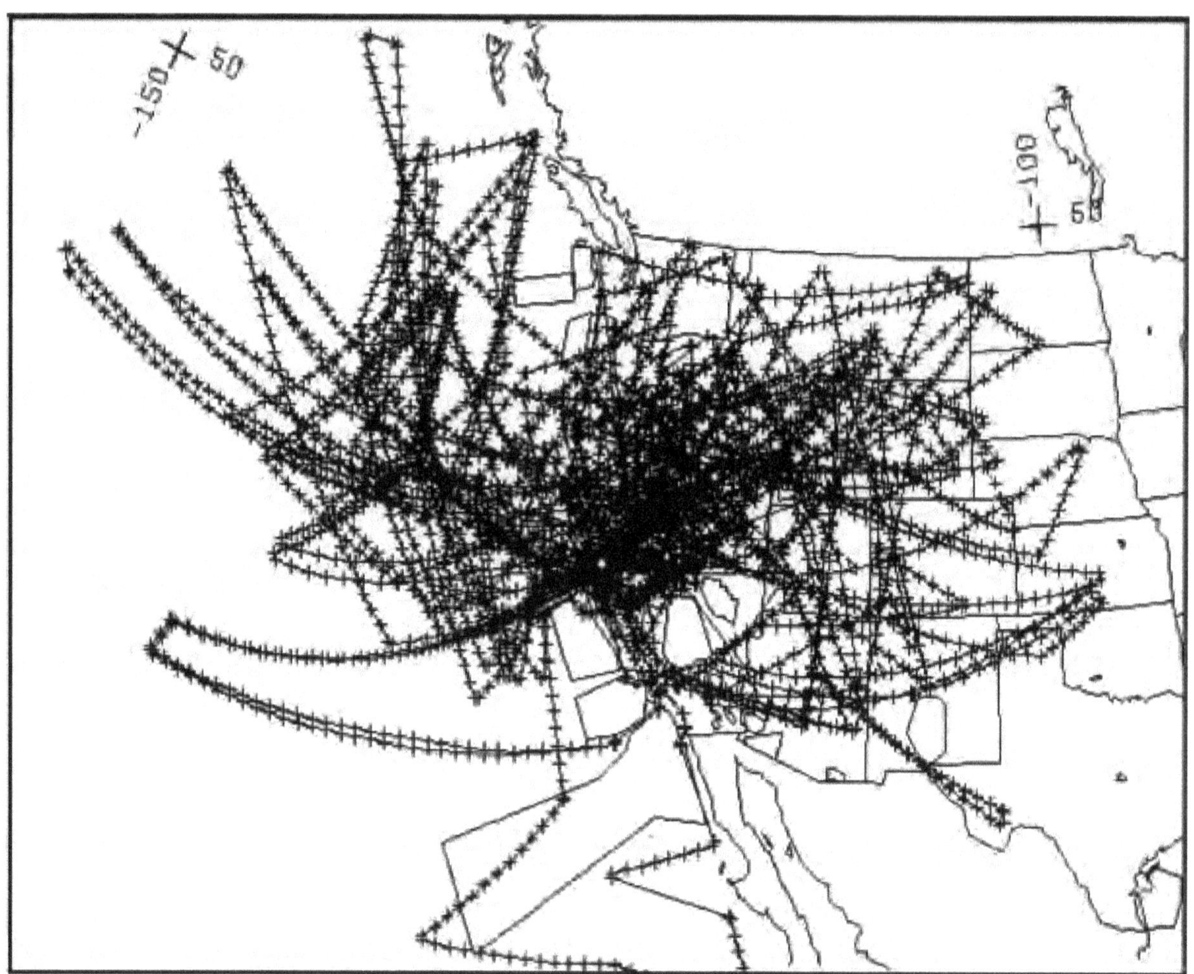

Figure 12. Superposition of 48 KAO astronomy flight plans for missions originating at NASA Ames during 1993–1995. Note the avoidance of SUA areas shown in figure 10.

Typically the aircraft would reach FL350 (flight level of 35,000 feet) half an hour after takeoff. Then the cavity door would be opened and the instrument and telescope systems checked while tracking on a conveniently located star. Climbing continued as fuel burned off; most observations were made at FL410. Usually "sky-noise" at lower altitudes was a problem, at least for sensitive instruments working in the thermal infrared. This was due to the higher water vapor overburden at the lower altitudes, and possibly to outgassing of the telescope and cavity that decreased with flight duration. Climbing later in the flight to FL430 and then FL450 was done as fuel-load permitted if observers requested altitudes above FL410. At these higher altitudes, some crew members were required to wear their oxygen masks because of the short time of consciousness in case of a rapid decompression, so flying above 41,000 feet was not popular with the crew. Rarely, but when necessary for ephemeral events such as occultations, 10-hour or longer research flights were flown.

Adjustments to flight durations and profiles were made as needed to maximize success. For example, when the tropopause (see figure 4) was expected to be high, sometimes three 5-hour flights (requiring lighter fuel loads) were flown instead of two 7.5-hour flights so that more time would be available at or above FL410.

4.5.2 Preparations for Flight

Hands-on in-flight and between-flight access to the aircraft, telescope, science instruments, and support equipment are major strengths of airborne astronomy. For example, in the daytime between flights, science instruments, mission systems, and the aircraft were serviced, discrepancies noted in flight were treated, and all but major system upgrades were implemented. The immediate availability of experienced staff for these activities was critical to effective operation. In the morning following a flight, and before starting work, the day crew would review the flight log and the items needing attention ("squawks") written on a white-board in the hangar next to the plane by the Mission Director. Short pilot-proficiency flights were sometimes flown on mornings after research flights when there were no critical systems issues, while personnel who had flown were sleeping.

Typically science-instrument installation began in the morning of the day before the first flight, although sometimes on the day of the first flight, by experienced instrument teams with mature instruments. There was usually no work on weekends, except by scientists completing flight preparations in the instrument laboratory and/or replenishing cryogens on their instrument (which may already have been installed on the telescope).

A science instrument was installed and its basic operation on the telescope was verified by the instrument team, with assistance from the observatory staff as needed. This included mechanical attachment, balancing, and optical alignment optimization on the telescope, verification of the instrument control and readout electronics, and communication with the observatory data system.

Three specialized light sources were developed to check alignment and operation of the instruments. Adjustment of the optical alignment of the instrument to view the secondary mirror of the telescope properly was done with a "Chopped Hot Plate" (about the size of the 7-inch-diameter secondary mirror) installed on the headring of the telescope in front of the secondary mirror. Then the hot plate was replaced by a "Portable Chopped Light Source" (PCLS, aka "Erickson Source") to establish the "boresight" between the instrument and the focal plane imager. The PCLS was a 6-inch-diameter telescope with a hot quartz bulb that produced both infrared and visible-light focused images in the focal plane of the telescope.

Once the instrument was physically installed and optically aligned, and the telescope balanced, system noise checks could be done while the telescope was operating. For example, low-level signals could be injected into the telescope's servo motors to check for microphonics and magnetic pickup in the instrument's detector electronics.

Often the "Arizona Collimator," a 1-meter-diameter telescope, was mounted on the fuselage exterior (fig. 13) to produce focused infrared and visible images in the KAO focal plane. This enabled "practice flights"—simulations of in-flight data acquisition done on the ground with all systems operating (including the telescope stabilization and chopping secondary mirror, but excluding the aircraft)—prior to the first flight of a series. Although excellent aids for installing science instruments, these light sources required access into and work inside the telescope cavity.

This was undesirable because of the possible introduction of dirt on the telescope mirrors and disturbance of telescope systems.

The KAO was often operated at night on the ramp in front of its hangar to verify and adjust the telescope systems by tracking on stars. Usually these "line ops" were done with a science instrument installed and members of the instrument team participating. These operations benefited from the generally mild climate and favorable meteorological conditions at Ames.

These procedures—used to verify a science instrument's operation during installation, and to adjust, debug, and verify the observatory's science mission systems on the ground with the instrument and telescope operating—contributed greatly to ensuring successful in-flight operation.

Figure 13. The KAO telescope installed in the C-141 (1990). Forward on the fuselage is left in this image. Optical alignment and verification of science instruments were done by mounting specialized light sources on the fuselage and/or on the telescope headring or spider.

4.5.3 Communication, Teamwork, and Co-location

Communication, the essence of teamwork, was a great strength of the KAO program. Stand-up status briefings for all interested parties were held daily after lunch. Preflight briefings for those flying were held about an hour before takeoff.

The contract staff and the civil service ground crew had offices in the KAO hangar (Ames building N-248) whereas flight crew personnel and civil service mission staff had their offices in the main hangar, N-211, as shown in figure 14. (Staffing details are given in section 4.7.) The proximity of these buildings and the presence of engineering offices, machine and electronics shops, and science laboratories all within half a mile of building N-248 at Ames were major assets enabling efficient support of the KAO. The civil service staff and support service contractors worked smoothly together. The close proximity of facilities and personnel made it straightforward to arrange meetings, enabled impromptu discussions, and fostered spontaneous brainstorming among members of the various staff elements, Ames support groups, and investigators. The formal and frequent informal encounters, and the shared concerns, engendered a spirit of teamwork that was extremely effective in resolving issues and solving problems. Thus, collocation was a profound factor in achieving the program's success.

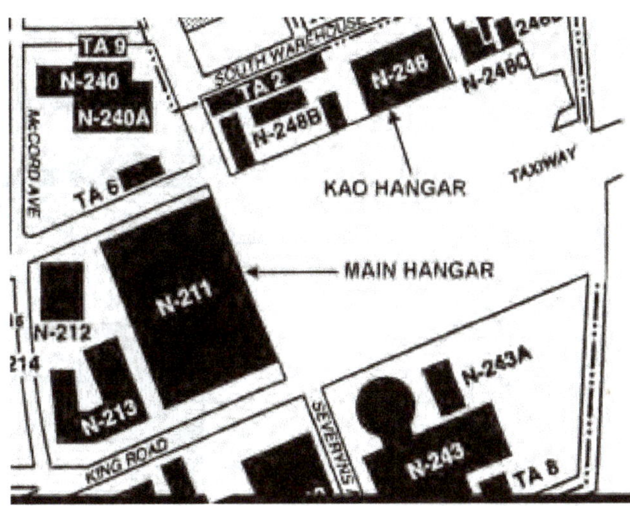

Figure 14. Configuration of research aircraft hangars at NASA Ames. The Main Hangar, building N-211, housed the ER-2, Learjet, and other research aircraft, as well as the Ames sheet metal shop where much of the hardware supporting the airborne science aircraft was fabricated. The Ames Engineering Division, where a number of observatory improvements were designed, was conveniently located in N-213, just behind the main hangar. The Space Science building with its labs and instrument machine shop was located about 200 yards above the KAO hanger as seen in this diagram.

4.5.4 In-Flight Personnel and Accommodations

On typical observing missions, a crew of seven operated the observatory. Of these, two pilots and a flight engineer in the cockpit were required to fly the aircraft. The pilot (fig. 15) was in command of the facility in flight. The mission crew and science team activities were coordinated by the Mission Director, who also handled all communications with the cockpit.

Figure 15. Piloting the KAO. Warren Hall, chief of the Science and Applications Aircraft Division (1994), is at the controls of the C-141.

Three experienced technical specialists operated the telescope and related systems. A Telescope Operator controlled the basic observatory systems, including the cavity door and telescope drive systems. A Tracker Operator controlled the video acquisition- and tracker-cameras on the telescope, acquired targets, and monitored pointing. Observatory computers were the responsibility of the ADAMS (Airborne Data and Management System) Operator.

On board the aircraft, personnel communications were facilitated by an intercom system with several configurable channels, and by the layout of consoles shown in figure 16. The latter allowed considerable nonverbal communication—eye contact and physical gestures—between mission staff and science team members, thereby reducing voice traffic on the intercom, which could be heavy in some circumstances.

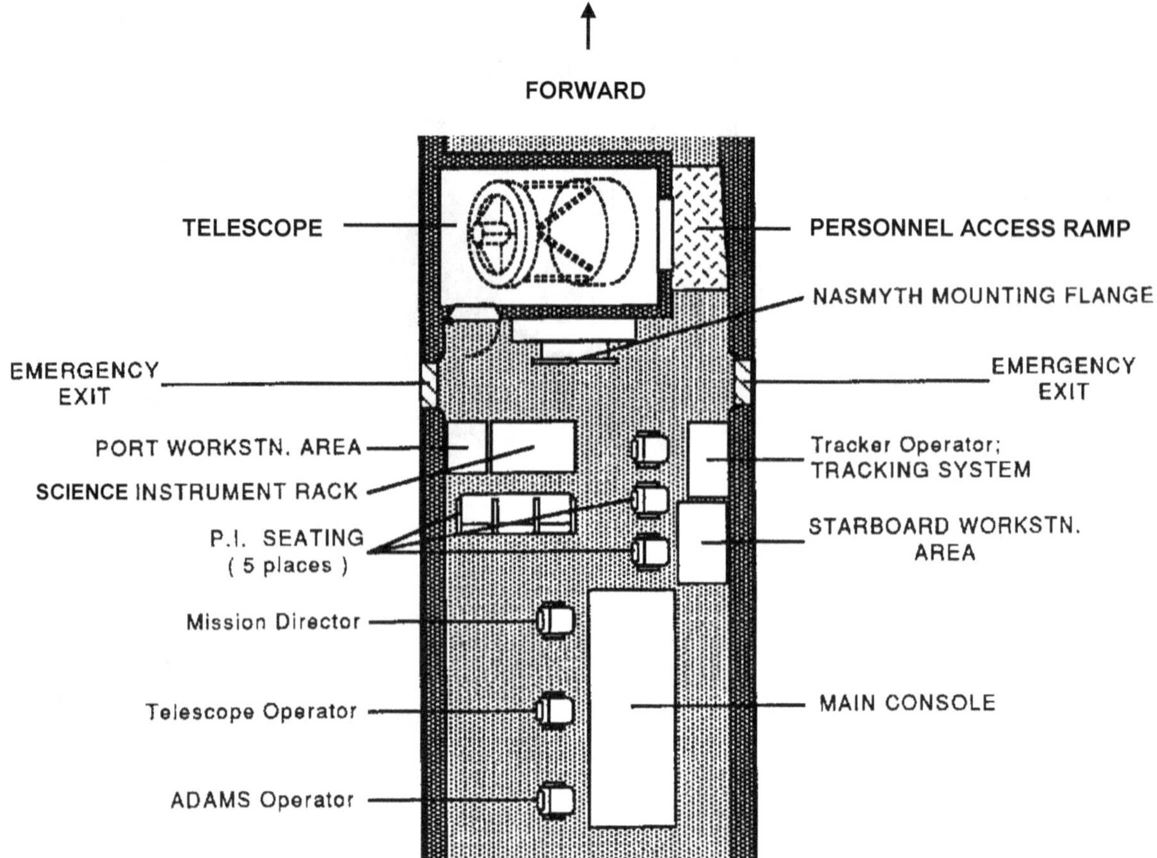

Figure 16. Layout of mission personnel accommodations on the KAO. The personnel access ramp behind the telescope enclosure led to the cockpit, the galley, and the toilet. Aft of the area shown were the observatory computer racks, the coffee and lunch area, spare equipment, seating for additional personnel, air-bearing compressors, and liquid nitrogen tanks. The interior width of the fuselage was 10 feet 3 inches (3.1 meters).

Monitoring and control of the observatory systems evolved over the life of the KAO, finally resulting in 11 simultaneous (and 23 optional) screens of video displays of aircraft and mission-systems parameters (ref. 18). See figure 17.

The Mission Director, Telescope Operator, and ADAMS Operator occupied adjacent seats at the Main Console (figs. 16, 17, and 18), which facilitated communications and occasional swapping of roles as needed. Role-swapping was possible during routine data-taking flight legs because of the experience of the mission crew members, and reduced the need for redundant crew.

Figure 17. The Main (telescope) Console (1995). This view, looking forward toward the starboard side of the C-141, shows the operating stations (from right to left) of the Computer Operator, the Telescope Operator, and the Mission Director. Beyond the Main Console at the far left is the Tracking System. The many real-time video displays were a major contribution to operating efficiency, as was the layout of the consoles, which enabled effective nonverbal communication.

The tracking station was closer to the science-instrument electronics rack, which was appropriate because of the need for nearly continuous interaction between the scientists and the Tracker Operator (figs. 16, 17, and 19). So in all, the minimum observatory staff (aft of the cockpit) totaled four people. It can be argued that automated systems could have been developed to reduce this number. However, it seems unlikely that a smaller staff would achieve the cost-effectiveness and high success rate enabled by the combined experience, flexibility, and rapid response capability of the competent four-person mission teams.

Figure 18. Operation of the telescope in flight. This picture shows the Main (telescope) Console looking aft. Jim McClenahan (standing), the KAO Facility Manager, is operating the telescope. Lou Haughney, then the KAO Project Manager and Chief of the Airborne Astronomy Branch, is serving as Mission Director. It was common for the top-level KAO managers to fly as Mission Directors; some qualified also as Telescope Operators.

Science teams consisted of two to perhaps eight members, depending on the instrument and the number of participating graduate students, postdoctoral associates, and Guest Investigators (GIs) whose programs were under way. Thus about a dozen people—three air crew, four mission crew, and five scientists—typically flew on astronomy flights.

Additional observatory staff often flew for training and occasionally as needed on observing flights to analyze and treat unusual problems with science mission systems. When practical, observations were made while in transit to and from deployments at remote sites, so a few ground-crew members or other staff would be carried. During the last few years of the program, pairs of teachers participating with science-instrument teams (see section 4.6) and their mentor also flew. Occasionally media representatives and others flew too. On occasion, then, the number of souls on board astronomy flights was as high as around 20.

4.5.5 In-Flight Flexibility

The limited number of observing hours per year demanded maximization of data return on each flight. Flexibility to alter observing plans in flight enabled this capability on many occasions when the science return would otherwise have been reduced due to unforeseen conditions. The latter included deviations from predicted high-altitude weather (typically winds) and air traffic, equipment malfunction, unexpected science results, and (rarely) denial of normally available restricted airspace. Such exigencies often required alteration of the flight plan, ranging from minor adjustments to the duration of planned flight legs to generation of a new flight plan for the remainder of the mission.

Figure 19. The tracking system. Displays of the three telescope cameras (Acquisition, Tracker, and Focal Plane), and fine slewing control enabled the Tracker Operator to acquire and track objects with the telescope. Seen here are Professor Patrick Thaddeus, an observer from Columbia University, and Allan Meyer, the Chief Tracker Operator. Thaddeus is verifying telescope pointing with a transparent overlay of the star field on the Tracker Camera video display. Investigators often provided some relief for the Tracker Operator by monitoring or adjusting the telescope pointing during data acquisition (see figure 20). Meyer also supported observers by assisting with flight planning and compiling much useful information on objects observed, etc. In addition to his routine duties, the Tracker Operator had a major responsibility for significant in-flight modification of the flight plan when necessary.

Small perturbations to the flight plan were often granted when a short increase in integration time on a particular object would significantly enhance the science obtained. For example, short legs were typical for bright calibration objects. If data acquisition on a calibrator could not be completed as scheduled (for whatever reason), the data on the primary science object(s) comprising most of the flight time could be compromised, so a short extension of the calibration leg could save an entire flight.

Significant changes to the flight plan usually entailed some loss in observing duration and required real-time approval from the responsible air traffic control authority. These events occurred on roughly one flight in seven, so the procedure for effecting them became routine in the sense that it was accomplished approximately 10 times per year. When major in-flight course replanning was required, it typically produced unexpected headings for the aircraft. For aborted flights, the revised plan was usually generated to return as quickly as possible to the operations base. On these otherwise "dead" legs, on-board software allowed candidate objects to be found to permit some useful science or system diagnosis to be done.

Sometimes remarkable science results being obtained prompted in-flight changes in the observing plan. For example, when unanticipated star-signal dropouts were observed prior to a stellar occultation by Uranus, the science team was granted an extension of the planned flight leg beyond emersion and the phenomenon was seen to reproduce, confirming the discovery of the rings of Uranus (ref. 19). Another example was when a first attempt to measure the previously undetected Si^+ line at a wavelength of 34.8 μm found the signal to be far stronger than theoretical predictions, a mapping strategy was devised in real time to produce an image of the emission during the remainder of the planned flight leg (ref. 20). Each of these events produced a significant scientific discovery from a single flight leg.

Acquisition of scheduled targets began with the aircraft turning onto the true heading that was pre-calculated in the flight plan. At the same time, the coarse elevation drive rotated the telescope assembly to the predicted initial elevation angle of the target. The Tracker Operator then used an analog electronic "Tweaker Box" to slew the telescope from a known star in a recognizable star pattern to point within about 0.5 degree of the target. This initial slew was guided by both a "dead reckoning" system based on aircraft heading and measured telescope attitude, and by visual identification of star fields on the 30-Hz low-light camera image displays.

Many sources had no visible counterpart, so accurate pointing could only be obtained by calculated offsets from pre-selected nearby field stars. Identification of source locations, mapping patterns, and background reference positions was done using transparent overlays that had been prepared directly from the Palomar and European Southern Observatory (ESO) Sky Surveys, the Lick Observatory Proper Motion Survey, and in later years, from the Hubble Space Telescope (HST) Guide Star Catalog. A transparent overlay was compared (fig. 19) with the video monitor display of the star field seen by the focal plane camera on the telescope. Differential offsetting by manually slewing the telescope reliably obtained accurate pointing on invisible sources.

Figure 20. Updating telescope pointing. Professor Eric Becklin (UCLA), a KAO PI, using the "Tweaker Box" to adjust telescope pointing during data acquisition.

However, the star-field recognition procedure would not work readily when a significant in-flight change in the flight plan required acquisition of an object in a relatively obscure star field for which no overlay had been prepared. An automated star-field recognition system that stored star fields using real-time images from the cameras would have been much preferable. Also, offsetting efficiency could have been improved somewhat with a semi-automated system.

Manual adjustment and monitoring of science-instrument parameters was routinely done by instrument team members in flight (fig. 21). Access to the instruments during data acquisition obviated remote control of parameters (e.g., filter selection), simplifying instrument development and enabling real-time adjustments as needed.

Many flights were completed with partially successful data acquisition despite malfunctioning science-instrument or mission systems, because on-board personnel were able to improvise workarounds in real time. For example, if telescope articulation (nodding) commanded by a science-instrument computer failed for some reason, manual nodding could be accomplished by the Tracker Operator in the "Learjet Mode" with data-taking coordinated by the instrument team via voice commands, resulting in reduced but acceptable efficiency.

Thus in-flight flexibility enabled by the presence of experienced on-board personnel was an essential feature in the success of the KAO.

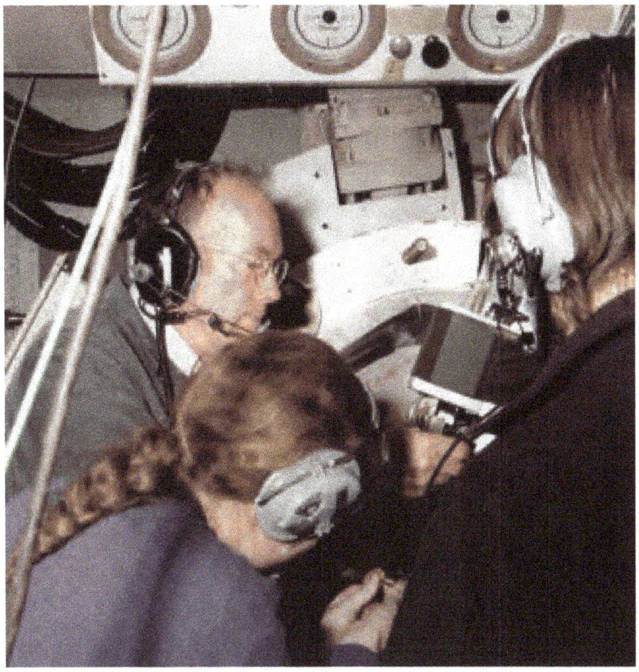

Figure 21. In-flight science-instrument adjustment. Professor Charles Townes (U.C. Berkeley) with graduate students Sara Beck and Donald Brandshaft make in-flight adjustments to their far-infrared Fabry-Perot spectrometer.

4.5.6 Make-Up Flights

The KAO management had a policy of providing a replacement (make-up) flight for an aborted or delayed flight, whenever possible, particularly if the problem was not due to a malfunction of the science instrument. Make-up flights occurred at most a few times a year. Typically, an additional flight was inserted in the schedule during the flight series when a flight was aborted, but sometimes later in the year, and rarely in the following year. This policy was intended to provide the approved allotment of flight time to investigators whose flights were aborted, and so was generally appreciated by them. Clearly, in some cases, the policy enhanced efficiency in obtaining data.

On the other hand, the resulting disruption of the planned flight schedule often imposed a hardship on the staff, on the investigators whose observations were delayed, and on their families. For example, staff members often had to alter their personal plans in order to fly on consecutive nights to support make-ups. The sacrifices of the staff to accommodate make-up flights are a tribute to their dedication. Alteration of the schedule to insert a make-up flight could compact the subsequent schedule, forcing the next investigator to hurry his installation, and/or fly with less rest, which may have reduced observing efficiency somewhat.

4.5.7 Investigator Facilities

Laboratory and office facilities provided by the KAO for visiting investigators were minimal. Only one dedicated laboratory area was available for the science-instrument teams. This area was frequently crowded because there were often two and occasionally three teams on site, so three labs would have been ideal. There were no high-fidelity telescope alignment or mission-computer simulators, so that instrument integration could only be done on board the aircraft, usually using the Ames-provided equipment, for example alignment light sources.

However, visiting KAO instrument teams had access to a variety of additional technical facilities and services at Ames on an informal basis. These included the Astrophysics Branch laboratories and machine shop (where instruments for airborne atmospheric and astronomical research were built), the Ames main machine and sheet-metal shops, and the laboratory facilities of the Astronomical Infrared Detector Group at Ames. The latter group, which developed and tested detectors for NASA astronomy missions, provided assistance ranging from consultation on sophisticated detector issues to lending liquid helium when deliveries to the KAO were delayed. The location of the base of operations for the KAO at a center that actively supported science-instrument and detector developments for ground-based, airborne and spaceborne astronomy missions was a significant benefit for the KAO operation.

Office space for investigators in the KAO hangar consisted of one small room adjacent to the laboratory in the hangar. For rest prior to flights, a couple of cots were provided in a storage room in the hangar. Toward the end of the program, Internet access was made available in the hangar, with computer assistance provided by the facility's computer staff whose offices were also located there. The Ames main library and the library in Space Science building N-245 stocked a wide variety of astronomical publications, which were also available to visitors.

Access to Ames required all visitors to obtain a temporary identification badge from the Ames security office during business hours. This usually worked smoothly, even for foreign visitors, if badge applications were submitted sufficiently in advance. When the KAO was on the ground, access to it was typically not available on weekends, or from 11 p.m. until 6 a.m. on weekdays.

4.5.8 Science-Instrument Airworthiness Approval

Ensuring safety was the purpose of NASA's process for authorizing science instruments to operate on aircraft. Efficient and straightforward procedures had been developed for the airborne science program at Ames for Earth sciences, atmospheric sciences, and astronomy research instruments prior to the KAO operation. The process was not related to the functionality or reliability of the instruments, which was admirably assured by the science teams (table 3).

In the early 1980s, larger, new KAO instruments were being developed, which increased safety concerns regarding, for example, mechanical attachments and cryogen volume. Assembly drawings were reviewed and approved in under a calendar week by the Flight Safety Office (FSO), with assistance from the Ames engineering organization when needed. The principal concerns of the FSO were the weight and moments of inertia, the fasteners, the vacuum-window seal design and construction of cryostats, and precautions for high-voltage if there was any. Commercial electronics were never questioned. After discussing the design with a cognizant instrument team member for an hour or so, any concerns were addressed, and final approval was given when these were met. For smaller, new instruments, drawings were not reviewed; instead, one of the staff would look at the instrument when it arrived and discuss its installation with an instrument team member. In some cases, a brief engineering analysis was done to ensure mechanical integrity. On occasion, instrument mounting hardware was designed and built by Ames personnel for visiting instrument teams. Authorization for installation required an aircraft work order, which normally included a hand-drawn sketch of the installation with pertinent parameters (weight, approximate center of mass, fastener details, etc.).

On subsequent flight series for the instrument, any significant changes to it were noted, and a new aircraft work order generated authorizing installation. Before each flight an airworthiness inspector always examined the installation, with particular attention to the attachments of the instrument to the telescope and the security of electronics in the investigators' rack.

The airworthiness approval process was so efficient as to be nearly transparent to the instrument teams. There were no significant safety incidents relating to science instruments over the 21 years of astronomy flights on the KAO.

4.5.9 Computers and Software

Applications of computer technology on the KAO over its lifetime expanded with the explosion of available computing capabilities during the same period, and were a major factor in increasing observatory effectiveness. Software development, debugging, and upgrades were greatly simplified by wide use of standard consumer-level software protocols, security procedures, applications, and hardware on the science-instruments and mission systems. Use of commercial components held down the cost of spares for off-line or in-flight swaps. The on-board data-system architecture as it had evolved by 1995 is shown in figure 22.

Of course there was a great deal of custom software. Its development efficiently accommodated desired improvements in functionality and flexibility as perceived, unencumbered by formal NASA or Federal Aviation Administration (FAA) requirements. Had it been subject to FAA standards for certifying avionics software (RTCA DO-178B, ref. 21), the science mission systems and instrument computer codes would have qualified under a Level D failure category: "Minor [impact] – Failure is noticeable, but has a lesser impact than a Major Failure (for example, causing passenger inconvenience or a routine flight plan change)," or Level E: "No effect [on passengers or flight plan]."

This level of software failure-immunity was acceptable because systems controlled by software were protected by manual and/or hardware provisions. For example, the cavity-door position was computer-controlled to track the elevation angle of the telescope (whether inertially stabilized or slewed under manual control); the door had hard-wired limits and physical hard stops, and its computer command-link could be interrupted or the door drive system disabled by the Telescope Operator, so software failure modes were not an issue. Similarly, the telescope attitude link to the aircraft autopilot that adjusted the aircraft heading (item 1 in table 5) was simply structured to be insensitive to computer error, and provided manual override control by the pilot. Total failure of an essential computer function would, of course, result in an aborted flight, requiring a flight-plan change, but this was routinely accomplished when necessary as described previously (section 4.5.1).

As a result of this approach, most mission-system and science-instrument software could be conveniently updated, between and even during flights, without disturbing the observing schedule. Minor changes to software (for example, in the tracker program) were sometimes tried on astronomy research flights without any previous testing; if the new version did not work, the previous version was loaded and used in flight, while modifications continued on the new version. This efficient and relatively informal approach was possible because the in-flight computer operators were often also system programmers, and because minimal approvals for software revisions were required. The computer team was proud of its many years of science-flight support performed without critical data loss due to computer support issues.

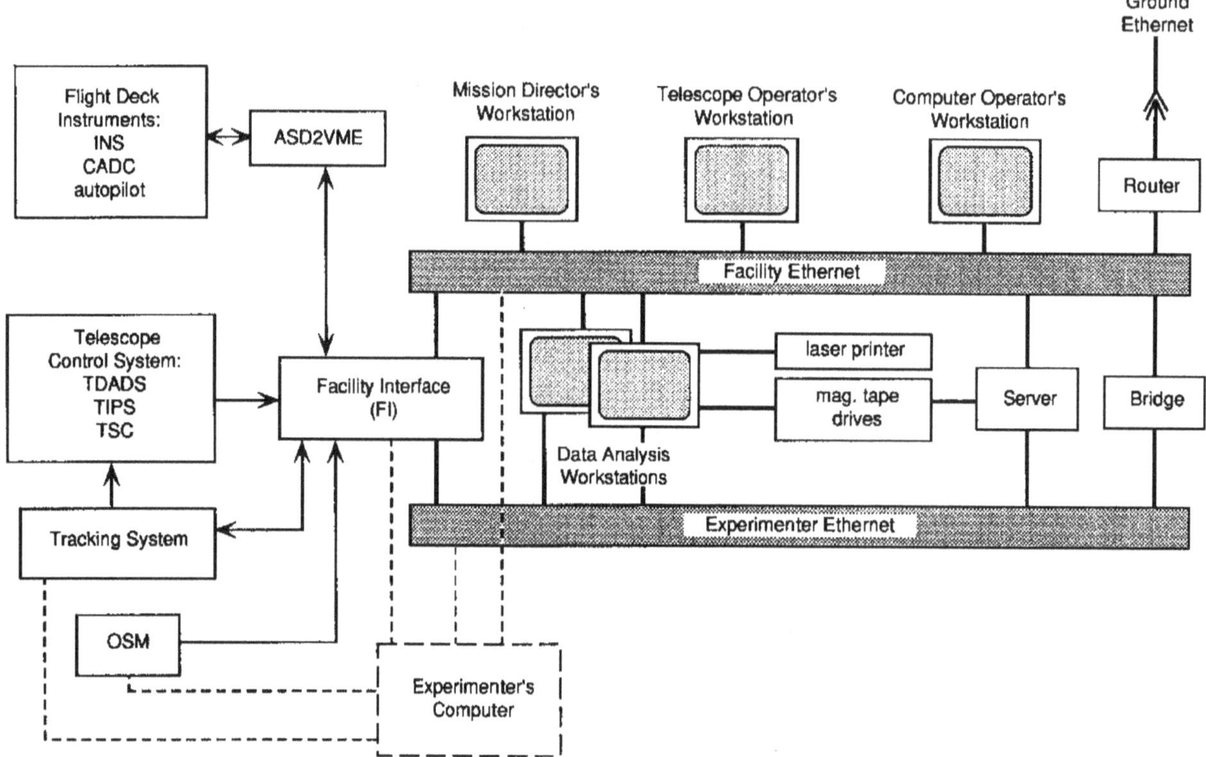

Figure 22. KAO data and control system architecture. Acronyms are INS: Inertial Navigation System; CADC: Central Air Data Computer; TDADS: Telescope Data Acquisition and Display System; TIPS: Telescope Inertial Pointing System; TSC: Telescope System Controller; OSM: Oscillating Secondary Mirror; and ASD2VME: Avionics Serial Data to VersaModular Eurocard inteface. Each science-instrument team provided its own "Experimenter's Computer," which interfaced to the observatory's data system as shown. Investigators used the observatory's Data Analysis Workstations to examine and process their data in flight.

4.5.10 Continuous Improvement: Hardware, Software, and Procedures

Beyond routine operations, the staff and scientists at Ames exploited the ongoing availability of the observatory to successfully orchestrate and implement numerous, significant systems upgrades throughout the KAO's lifetime. Additional support for these improvements was provided by the Ames mechanical, electronics, instrumentation, and aeronautical engineering groups, by industrial firms, and by KAO users from other institutions.

Some of the many technical innovations achieved are listed in table 5. Items 14–20 were provided by the NASA Facility Scientist. Most of the improvements were accomplished with multiple-year efforts in parallel with the flight program.

Incremental implementation of improvements on the observatory was often done on research flights, minimizing the need for separate engineering flights. As seen in table A1, over the program's lifetime, only 122 engineering flights of 4½-hour average duration were flown out of a total of 2,289 flights. A few of the upgrades, notably items (3) and (11) in table 5, required considerable downtime of the observatory.

Table 5. Some Technical Improvements to the KAO During Its Lifetime
1. Automated aircraft heading turner: keeps telescope centered in cross-elevation range
2. Command of basic telescope functions by science instruments
3. Improved passive flow control (PFC), based on SOFIA wind tunnel tests
4. Effective flight planning software, including in-flight route-modification capability
5. Upgrades to video tracking and acquisition cameras, and software
6. Dead-reckoning telescope pointing system
7. New telescope pointing control system
8. Upgrades to the cavity-door drive mechanism and control
9. Modernizations of on-board computer systems
10. Computerized mission-system monitors and fault annunciators
11. Upgraded cavity-environment control system
12. Upgraded chopping secondary mirror mechanisms
13. Optical-quality fixed secondary mirror
14. Optical alignment equipment and procedures for installing science instruments
15. Button mirror on telescope secondary mirror
16. Telescope infrared baffle plate
17. Optical-path purging system
18. Facility charge-coupled device (CCD) focal plane imager
19. Convenient software for calculating atmospheric transmission
20. Hardware for mounting a science instrument on the telescope

4.5.11 Local Infrastructure

Based at NASA Ames Research Center, the KAO operated from Silicon Valley in the heart of the San Francisco Bay Area—within a short drive from its three major cities. The proximity of commercial vendors, services, and airports at or near Ames contributed to the success of the KAO program. Local lodging and food service accommodations for visiting teams were conveniently close. Typically, within 10 minutes of leaving the KAO hangar after a night's flight, astronomers could be at their motel for their day's rest. The numerous, nearby technical parts' suppliers and Ames scientists' laboratories and machine shop (within 200 yards of the KAO hangar) were resources often exploited when mission systems and science instruments needed repair or maintenance to meet the flight schedule. The three international airports within 35 miles of Ames facilitated science-team travel for KAO flight series. A number of nearby universities (Stanford, U.C. Berkeley, U.C. Santa Cruz, U.C. San Francisco, and San Jose State) provided an ample and accessible academic environment. The recoating of the KAO telescope mirrors was accomplished at Lick Observatory, a multi-campus research unit of the University of California, on Mount Hamilton near San Jose.

4.6 EDUCATION AND PUBLIC OUTREACH (E&PO)

Throughout the KAO program, publicity for noteworthy scientific results was handled by the Public Affairs Offices at Headquarters and Ames in the usual manner for all of NASA's science programs. In addition, however, the KAO was a mobile ambassador, welcoming the public on board this operational NASA observatory on many occasions between flights. For example, thousands of people from the San Francisco Bay Area visited the KAO at the annual air shows at Moffett Field. When on deployments, the KAO staff would arrange observatory tours for the local residents. These events usually attracted hundreds of people interested in astronomy, aircraft, and NASA in general. These opportunities enabled visitors to interact with NASA personnel while examining this unique astronomical facility—a positive and stimulating experience.

In 1991 Dr. Dan Lester, a KAO investigator from the University of Texas, initiated a program of "Science in the Stratosphere" that enabled elementary and middle school science teachers to participate in astronomy research on the KAO. Two teachers participated each year, flying on three or four flights until the KAO was decommissioned in 1995. They observed and assisted scientists with their investigations, returning to their schools with expanded awareness of, and appreciation for, science in general and airborne astronomy in particular.

In 1993 the Flight Opportunities for Science Teacher Enrichment (FOSTER) program was initiated by Dr. David Koch, a research scientist in the Ames Astrophysics Branch, with support from an education grant from NASA's Office of Space Sciences. Ms. Edna DeVore, a former teacher and an administrative assistant, managed the program with assistance from the Ames Public Affairs Office. Competitively selected teachers participated in 1- to 2-week group workshops at Ames, with a variety of educational aspects including lessons on physics, astronomy, meteorology, and aeronautics led by NASA scientists, engineers, pilots, and the KAO staff. Subsequently, pairs of teachers became involved with KAO investigator teams, and then participated in one or two science flights with them (fig. 23). Benefits for the teachers included firsthand participation in a science investigation, establishing rapport with other participating teachers, development of curricula for their school districts, and inspiration from these experiences. Some 70 teachers participated during the 3-year FOSTER program. It was considered extremely valuable by the teachers (ref. 22).

Some of the teachers who flew on the KAO in both the Science in the Stratosphere and FOSTER programs helped to promote approval of SOFIA by informing NASA officials and key congressional staffers of their experiences and the E&PO potential of airborne astronomy.

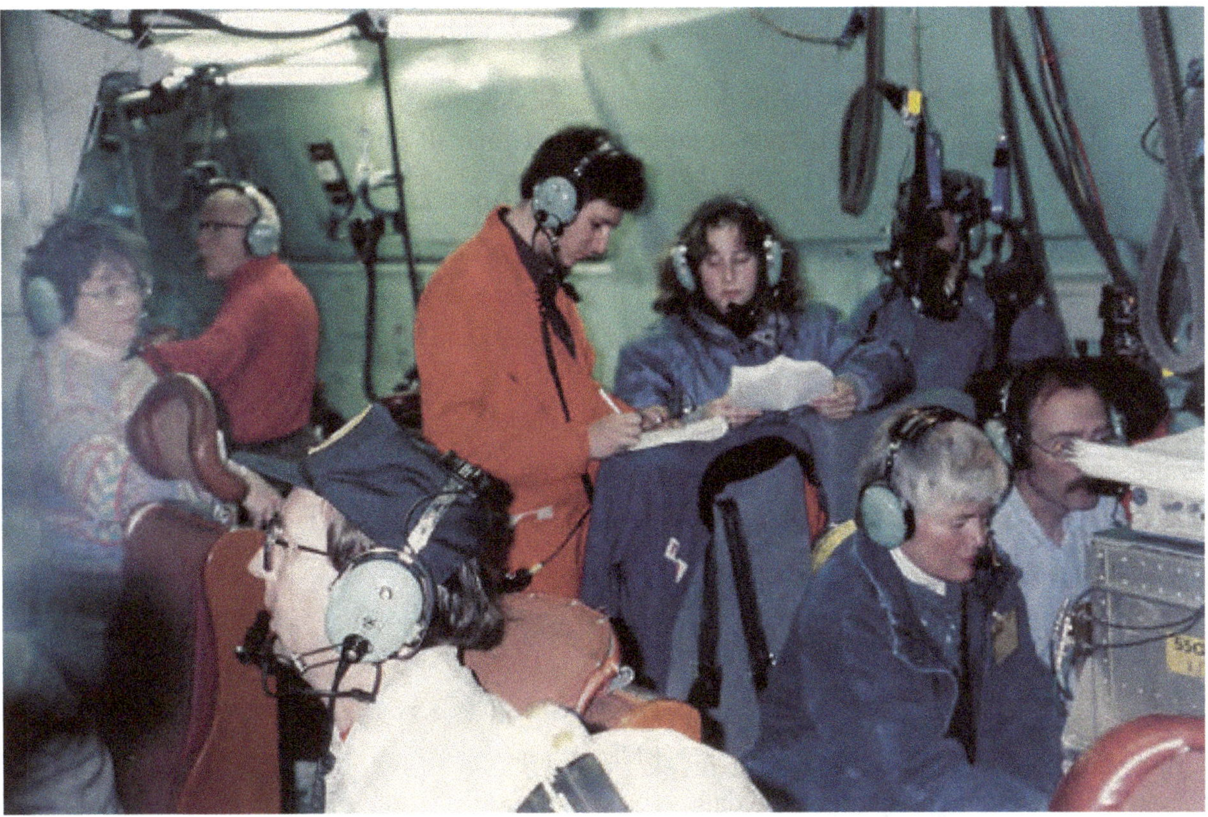

Figure 23. E&PO in action on the KAO. This photo was taken during data acquisition on a flight of the NASA Ames Cryogenic (Echelle) Grating Spectrometer (CGS). Personnel from left to right: Edna Devore, FOSTER program director; Carl Gillespie (in red shirt), project Science Coordinator who coined the acronym SOFIA; Dr. Sean Colgan, CGS team member; Dr. Alex Rudolf (in orange flight suit), National Research Council Associate working with the CGS team; Elizabeth Mason, reporter for the San Jose Mercury News; Penny Moore, physics teacher at Piedmont High School in California and originator of three national science education programs; and Dr. Mike Haas, CGS Instrument Scientist.

4.7 ORGANIZATION / STAFFING

4.7.1 General

Over the lifetime of the program, the organization of KAO operations evolved somewhat. The organization described here and depicted in figure 24 represents the later years of the program (ref. 23). At NASA Headquarters, the head of the Infrared Astronomy Branch served as both the Program Executive and Program Scientist. At Ames, the Flight Operations Directorate was responsible for all aircraft activities and the distribution of their resources, including those of the KAO program; there was no full-time program manager. Individuals in the different organizations shown in figure 24 reported to their supervisors in those separate organizations. All participants' efforts were coordinated by the management of the Airborne Astronomy Branch.

The operation was carried out by a mix of 26 NASA civil service (CS) personnel and 31 support service contractors, as shown in table 6. These are full-time equivalent (FTE) numbers, reflecting the fact that a minor part of the effort was supplied part-time personnel. Two basic categories of the activity were (1) operation of the aircraft, which was staffed almost entirely by civil servants, and (2) "mission operations," which was everything else. The effort for support of the aircraft operation was about half that for mission operations.

The CS staff was responsible for management of the program, for science support related to users' science instruments, and for some in-flight operation of the telescope and mission systems. The 31 FTE support-service contractors participated in in-flight mission-systems operations, provided nearly all of the data-systems support, and contributed to telescope maintenance and upgrades.

Table 6. KAO Operations Staff (FTE) at Ames

Activity	CS	Non-CS	Total	%
Mission Operations			31	55
Data Systems		7.5		
Telescope Systems	2	8		
Engineering		9		
Logistics		1		
Contract Administration		3.5		
Aircraft Operations			16	28
Airworthiness	3			
Flight Crew & Operations	4.5	1.5		
Maintenance & Servicing	7			
Management	7		7	12
Science Support	2		2	4
Contract Admin., SR & QA	0.5	0.1	0.6	1
Total	26	30.6	56.6	100

Table 6 and figure 24 include only organization and staffing for observatory operations. In addition there was on average an estimated (ref. 24) 5.5 FTE personnel from the resident staff in the Ames Engineering Directorate who provided mechanical, electrical, instrumentation, aeronautical engineering, and shop effort in support of observatory improvements, including those shown in table 5. Also not included in the operations description is the effort for E&PO, which probably averaged about 0.5 FTE over the KAO's lifetime. These efforts bring the total CS complement to about 32 FTE and the total manpower to approximately 62.5 FTE. Further details of responsibilities are provided in the following section.

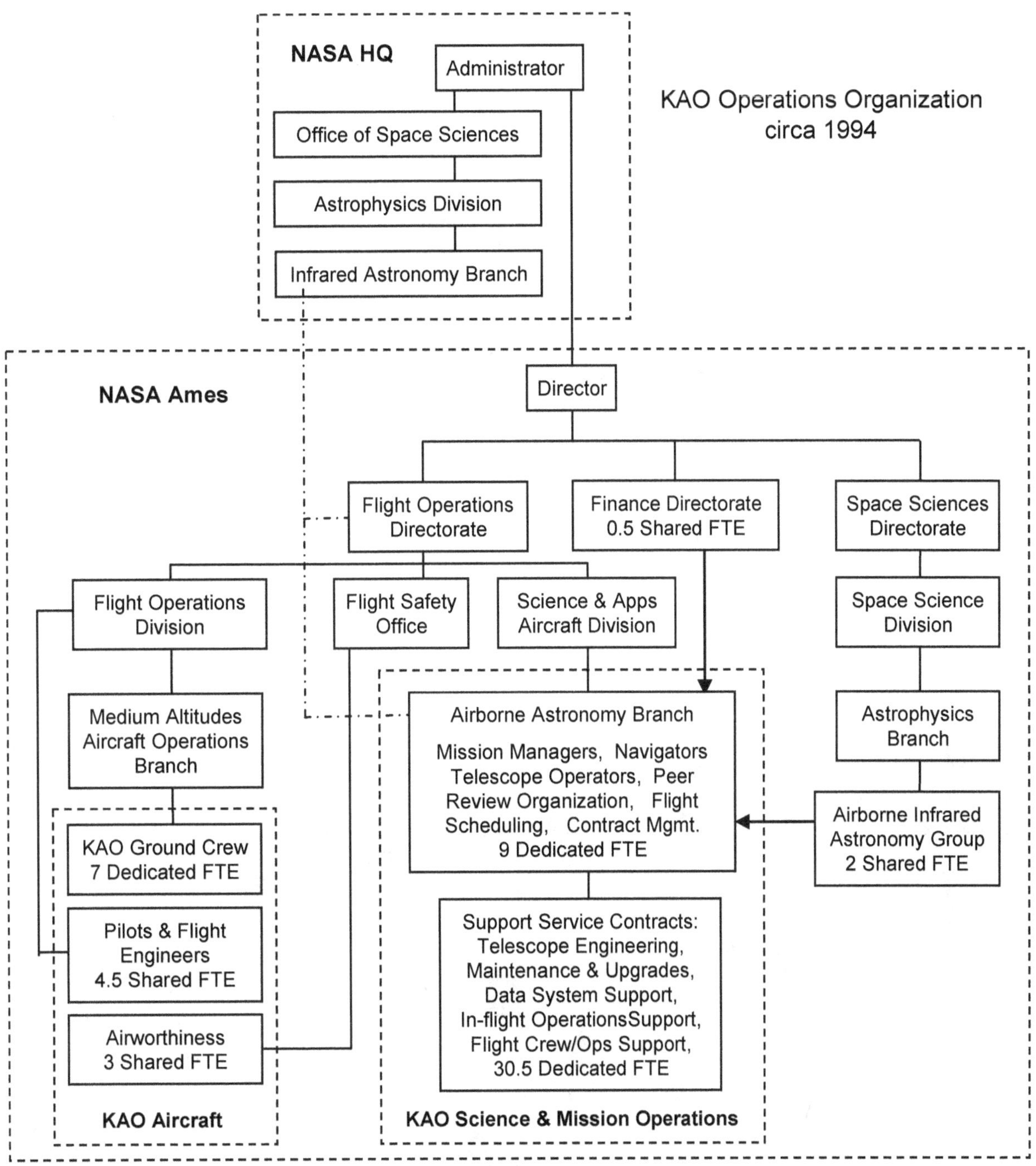

Figure 24. Organization of the KAO operation at NASA Ames. Arrows indicate continuing contributions on an as-needed basis.

4.7.2 Responsibilities

4.7.2.1 Aircraft

The KAO C-141 aircraft—registration ("tail number") N714NA—was the responsibility of the Aircraft Operations Division. A dedicated seven-man CS ground crew serviced and maintained the plane. The U.S. Air Force (USAF) operated about 285 C-141s, and the KAO was able to participate in their maintenance program. Medium maintenance was done at nearby Travis Air Force Base, where the KAO went 1 week per year for scheduled servicing, and as needed for unexpected issues. The Air Force cycled its C-141s and the KAO through its heavy maintenance depot at Lockheed in Marietta, Georgia, requiring 1 month every 5 years. Unanticipated significant maintenance while on deployment was also done with the assistance of the USAF; for example, replacement of an engine that failed while the observatory was based in Christchurch, New Zealand, depicted in figure 25. Efficient and economical aircraft maintenance was a major factor in the success of the KAO program.

KAO flight crew personnel were matrixed and flew other aircraft; qualified pilots from the USAF and from other NASA centers often flew the KAO. The Ames Flight Safety Office (FSO), with three FTE personnel, ensured the airworthiness of the plane, the mission systems, and the science-instrument installations. Authorization for departure was the responsibility of the FSO and the flight crew.

Figure 25. C-141 engine replacement. In 1988 a failed C-141 engine was replaced by the USAF in Christchurch, New Zealand, where the KAO was deployed for observations of southern hemisphere sources. The aircraft maintenance arrangement with the USAF was a major benefit to the KAO program.

4.7.2.2 Mission Systems

The KAO science mission systems, comprising the telescope, cavity door, computers, and ancillary systems, were the responsibility of the Airborne Astronomy Branch. Its operations and support people were considerably matrixed among KAO tasks to achieve efficiency and backup capability in operations, maintenance, and improvement of the mission systems over the lifetime of the program. For example, Telescope Operators participated in servicing the telescope systems, and Mission Directors participated in scheduling observations and arranging logistics for deployments. Nearly all the project staff took turns with in-flight responsibilities. Key project staff positions and individuals filling them in 1979 are listed in table 7. All except the Project Pilot and Facility Scientist worked full-time on the project. All but three were civil servants. It is a credit to the program and these people that nearly all continued in these roles until they either retired or until the program was terminated in 1995.

Table 7. Key KAO Project Staff Positions (from at least 1979, except as noted)

Role	Individual
Project Manager	Lou Haughney
Facility Manager	Jim McClenahan
Science Coordinator	Carl Gillespie
Senior Observatory Technician	Bruce Kelley
Senior Telescope Operator	Milo Reisner
Senior Observing Assistant	Allan Meyer
Records and Logistics Manager	Bob Barrow
Flight Planning Navigators	Jack Kroupa, Bob Morrison
Data System Manager	Tom Mathieson
C-141 Crew Chief	Lloyd Domier
Project Pilot	Warren Hall, Terry Rager (~1985)
Facility Scientist	Dr. Ed Erickson
Project Scientist	Dr. Ted Dunham (1990)
Deputy Project Manager	Wendy Whiting-Dolci (1992)

4.7.2.3 Science Support

In a practical sense, the entire KAO program was dedicated to science support, from helping to arrange peer reviews of proposals to operating the observatory for observations. Here however, we refer to the efforts by the Ames scientists and their groups in support of the observatory staff (fig. 24) and the approved investigator teams.

A scientist from the Ames Astrophysics Branch served as the KAO Facility Scientist throughout most of the program's lifetime. He and other personnel from the Ames science program furnished technical advice to the operations staff, organized the KAO users' group, recommended items and priorities for continuous improvement, contributed to science community outreach, and helped organize the Airborne Astronomy Symposia. The Ames scientists had a major role in advocating the program with managers at Ames and those in the Office of Space Sciences at NASA Headquarters.

With technical support from the Astrophysics Branch, the Facility Scientist carried out and/or supported some of the engineering tests of the observatory; for example, measurements of the image stability of the telescope, dependence of image size on infrared wavelength, and dependence of optical image quality on flight parameters (ref. 25). Likewise, he provided equipment for the telescope to improve science performance; to reduce spurious noise sources in infrared instruments, he included a black "backup" plate mounted behind the tertiary mirror and a button mirror mounted on the secondary mirror to reduce varying infrared background radiation, and a blower and plumbing to purge the light path leading to the science instruments.

Further, the Facility Scientist and his group provided direct support for KAO users. This included building optical equipment (chopped hot plate, portable chopped light source, targets, etc.) and establishing alignment procedures for installing instruments on the telescope. These were used extensively by most instrument teams, as was software furnished for flight planning (WINDO), for calculation of atmospheric transmission (ATRAN), and for automated in-flight bore-sighting of a science instrument with the telescope focal plane camera (PEAKR). Sophisticated, custom facility flight hardware was developed that was used with some visitors' instruments, for example the instrument mounting hardware and the facility CCD focal plane camera (ref. 26) shown in figure 26.

Figure 26. Science-instrument support. The Cornell University grating spectrometer (1995, see table 8) is shown installed on the telescope. The mounting hardware and the digital focal plane camera (black module with KAO logo) were provided by the NASA KAO Facility Scientist.

The facilities of the scientists at Ames were available to, and were often useful to visiting science teams. The experienced machinists in the Space Science Instrument Shop repaired and/or remade parts for visitors' instruments on an as-needed basis. The science-instrument laboratories were also resources for visitors. Loans of hardware for cryogen handling (funnels, dewars, transfer tubes, etc.) and liquid cryogens were common when found lacking in the visitors' lab at the KAO hangar. Leak checking, and vacuum hardware and systems were similarly available. Optical alignment equipment (lasers, mirrors, a theodolite) was borrowed on occasion, and measurements of infrared filter spectral response were made for some visitors using the available laboratory spectrometers. Also appreciated was use of the Space Science Library (see section 4.5.7).

Finally, the Facility Scientist and other members of the KAO instrument teams at Ames consulted with visiting scientists on observing techniques and problems, recent performance of observatory systems (e.g., the oscillating secondary mirror), and about science related to various astronomical objects of common interest.

In 1990, the position of KAO Project Scientist was established. It was also filled by a KAO instrument team leader from the Astrophysics Branch. He helped organize and select panelists for the annual peer reviews of observing and instrument proposals, in cooperation with the Chief of the Infrared Astronomy Branch at NASA Headquarters. Prior to 1990, the peer reviews were arranged by the staff of the Airborne Astronomy Branch.

Direct support for the KAO from the Astrophysics Branch amounted to about two full-time equivalent (FTE) people providing science, engineering, and technical effort (fig. 24 and table 6). Many of the contributions to the observatory from Ames scientists were the result of experiences with their own instruments and observing programs. Their enthusiasm and rapport with the observatory staff helped to energize the entire program.

4.8 SCIENCE PROGRAM

4.8.1 Science Instruments

Fundamental to the success of the KAO was the vigorous focal plane science-instrument development program it intentionally spawned in the science community. As a suborbital NASA facility, the KAO was expected to support the exploration of new instrument technologies and training of young scientists. Instrument teams were led by scientists from universities, government laboratories, and industrial concerns, both U.S. and foreign. They developed the instruments at their home institutions, installed them on the telescope (fig. 27), operated them in flight (fig. 28), and analyzed and published the data.

The science instruments typically comprised the most recently developed, high-tech equipment on the observatory. Instrument configurations were usually changed between flight series to achieve improvements. Despite these facts, the records (table 3) show that their reliability was higher than that of either the aircraft or the telescope. This was likely due in large part because the instruments were operated by their developers.

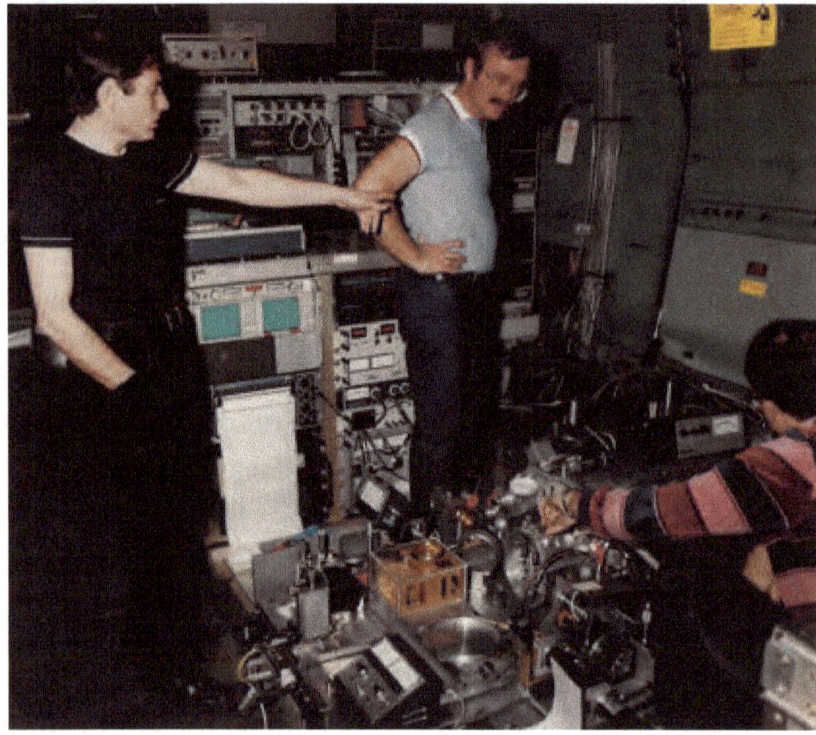

Figure 27. Installing a science instrument. Professor Hans-Peter Roeser (Max Planck Institute for Radio-Astronomy, Bonn, Germany) directs the installation of his heterodyne spectrometer on the KAO (1985). Dr. Peter van der Wal is at center and Dr. Roland Wattenbach is on the right.

About 50 specialized science instruments encompassing a wide variety of technologies and capabilities were developed and used by 33 different instrument teams during the KAO's lifetime. Of course in the early days of the program, infrared detector technology was primitive, and many of the instruments were correspondingly simple so that development could occur on a much shorter timescale than in the 1990s, or especially today. Nearly all the instruments from U.S. institutions were developed largely with KAO funding, awarded in grants (not contracts) on the basis of peer review. In many cases, additional resources from the developers' home institutions were applied to develop the instruments. Some of the instruments operated at wavelengths that permitted use (and so were used) on ground-based telescopes as well as on the KAO.

Sixteen of the instruments that existed in 1995 are listed in table 8. Not all of these were flown in that year. Only one of these, the NASA Ames Cryogenic (Echelle) Grating Spectrometer (CGS), approached the status of a "facility instrument," with which some Guest Investigators were provided data that they published without members of the instrument team as co-authors. In addition to the instrument team leaders shown in table 8, the other KAO instrument team leaders are indicated in tables B1 and B2.

In table 8, spectral resolution is given (a) for heterodyne receivers as $\delta\nu$, where ν is frequency, and (b) for instruments employing direct detectors as resolving power $\lambda / \delta\lambda$ where λ is wavelength. Frequency and wavelength are related by $\lambda = c / \nu$, where c is the speed of light.

Table 8. KAO Focal Plane Instruments Existing in 1995				
Principal Investigator/ Affiliation	**Instrument Type**	**Wavelength Range (μm)**	**Spectral/Spatial Channels**	**Spectral Resolution**
A. Betz / U. Colorado	Heterodyne Spectrometer	60–400	512/1	δν = 3 MHz
J. Bregman / NASA Ames & D. Rank / Lick Observatory	Photometer/Camera	2–5, 6–13	1/128x128	Various (Filters)
E. Dunham / NASA Ames	High Speed CCD Photometer/Camera	0.3–1.1	1/2048x2048	Various (Filters)
E. Erickson / NASA Ames	Echelle Spectrometer	16–210	32/1	λ/δλ ~ 1000–5000
D. Harper / Yerkes Observatory	Photometer/Camera	30–500	1/8x8	λ/δλ ~ 2–10
P. Harvey / UT Austin	High Angular Resolution Camera	40–200	1/2x10	λ/δλ ~ 20–100
T. Herter / Cornell U.	Grating Spectrometer	5–36	128/128	λ/δλ ~ 100–9000
R. Hildebrand / U. Chicago	Polarimeter	100	1/6x6	λ/δλ ~ 2.5
H. Moseley / NASA GSFC	Grating Spectrometer	16–150	48/1	λ/δλ ~ 35–200
H. Larson / U. Arizona	Michelson Interferometer	1–5	1	λ/δλ ~ 1000–300,000
H. Röser / DLR Berlin (DE)	Heterodyne Spectrometer	100–400	1400/2	δν ~ 1 MHz
R. Russell / Aerospace Corp.	Prism Spectrometer	2.9–13.5	58/1 & 58/1	λ/δλ ~ 25–120
G. Stacey / Cornell U.	Imaging Fabry-Perot Spectrometer	18–42	1/128x128	λ/δλ ~ 35–100
C. Townes / UC Berkeley & R. Genzel / MPE Garching, DE	Imaging Fabry-Perot Spectrometer	40–200	1/5x5	λ/δλ ~ 3000–300,000
F. Witteborn / NASA Ames	Grating Spectrometer	5–28	120/1	λ/δλ ~ 300–1000
J. Zmuidzinas / CalTech	SIS Heterodyne Spectrometer	370–600	160/1	δν ~ 0.6, 3.0 MHz

4.8.2 Investigators, Observing Allocations

Peer review panels awarded observing time in units of flights to Principal Investigators (PIs) who were the lead scientists on approved observing proposals. The PIs included instrument team leaders and Guest Investigators (GIs)—astronomers proposing to observe using others' KAO instruments.

Peer reviews took place annually, with additional reviews to accommodate unanticipated targets of opportunity, such as Supernova 1987A. Over the program's lifetime, there were 126 different PIs approved for about 510 investigations on 1,424 astronomy research flights (table 1), resulting in roughly 1,000 publications (not including talk abstracts, etc.) (refs. 1,2).

Figure 28. Data acquisition in flight. Drs. Larry Caroff and Ed Erickson (NASA Ames) examine the quality of incoming data (1979). This capability enabled prompt decisions to make best use of observing time. Caroff later moved to NASA Headquarters to manage the infrared programs in the Astrophysics Division. Erickson was the KAO Facility Scientist.

NASA provided financial support for PI teams from U.S. institutions. Expenses covered typically included some salaries, travel, equipment and software, consumables, publications charges, etc. The larger grants usually went to the instrument teams, to support both their instrument work and their observing programs.

All 126 PIs, and the number of flights awarded each, are listed in table B1. Data from this table were used in producing figure 29, which depicts the corresponding distribution of flights per PI.

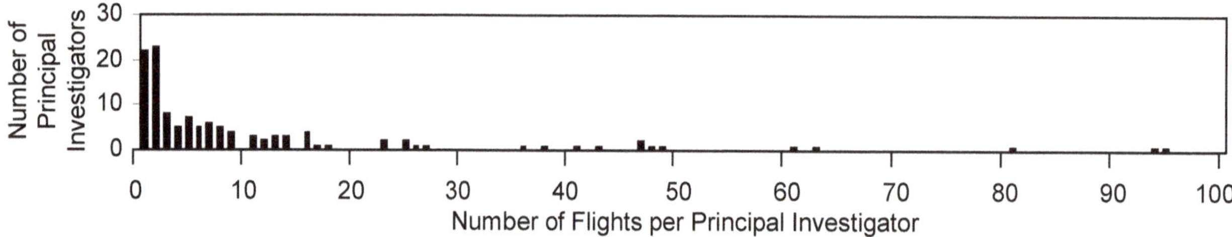

Figure 29. Distribution of astronomy flights awarded to Principal Investigators.

About half the flights were awarded to 18 of the PIs, nearly all of whom were instrument team leaders. GIs were encouraged to propose KAO observations in collaboration with members of any science-instrument team whose operational instrument was funded by NASA. The distribution of awarded flights, and the requirement (for nearly all the instruments) for GIs to collaborate with an instrument team on observations and publications, led to an impression among some members of the science community that the KAO was a "closed shop."

The situation evolved somewhat. Early in the program, instrument teams outnumbered GIs; by the end of the program, the number of GIs exceeded the number of approved instrument teams. For example, in 1995—a year in which there were only 51 research flights due to major aircraft maintenance—10 different instruments were flown supporting a total of 25 different PIs; 24 of the 51 flights were awarded to instrument teams. In 1994, 9 instruments were flown supporting 33 separate investigations carried out on 74 astronomy flights, of which 35 flights were for the instrument teams. Usually GI's observations occurred on flights shared with the instrument teams, which was efficient for both.

It is not surprising that the instrument teams were so successful in the competition for flights. These teams invested a major part of their research effort on KAO observing, and typically had strong science programs that motivated their observations and instrument developments. Also, the support for the instrument teams was partially justified by their important contributions of technology advancement and training of young instrumentalists.

4.8.3 Data Availability

Data taken on the KAO were the property of the PIs (fig. 30). Data were not archived, and no log of the data acquired was publicly available. Although certainly all the data were judged for quality and potential scientific value by the PI teams, valid data from some observations were never analyzed or published. This was due, at least in part, to the fact that most of the instrument teams focused much effort on their instruments, and so exploited effectively only those data they considered most interesting.

Early in the program, most instruments were evolving rapidly so that data analysis needed to be modified for each flight series. Even later in the program, when most instrument configurations were becoming more stable, most were still being changed between flight series, requiring adjustments to the data analysis. Archiving and publicly posting all valid data in calibrated, useful formats would have been possible with considerable additional effort (cost), perhaps increasing

Figure 30. After a flight series. Graduate students Jackie Davidson and Sean Casey, and Professor Al Harper (U. Chicago) with the Yerkes far-infrared photometer/camera (see tables 8, B1). Following a successful flight series, the instrument and data were taken to the home institution of the Principal Investigator.

scientific productivity. However, even as late as 1995, money was probably better spent on improving the instruments because available instrument technologies were still evolving rapidly.

4.8.4 Science Involvement at NASA Ames

Recognizing the potential of airborne astronomy as a discipline and the value of an in-house group for supporting and exploiting it, and for other astronomy research, Ames Director Dr. Hans Mark established the Astrophysics Branch at Ames in late 1970 (see also section 4.1). The scientists in the branch began a vigorous program of infrared instrumentation, which included a balloon-borne telescope and focal plane instruments for observing from both the Learjet and from ground-based telescopes. In addition, under the leadership of the Branch Chief Fred Witteborn, they initiated studies for a Shuttle Infrared Telescope (which subsequently morphed into the Space Infrared Telescope Facility, SIRTF, later renamed the Spitzer Space Telescope). Witteborn and his team are shown in figure 31.

Figure 31. A NASA Ames instrument team. Left to right: Drs. Dan Lester, Jesse Bregman, Harriet Dinerstein, Fred Witteborn, and Mr. Harold Crean (1980). Witteborn was the team leader and was the initiator of studies that led to the development of NASA's Spitzer Space Telescope. Lester and Dinerstein were postdoctoral associates. Lester later organized the KAO-related teacher education program "Science in the Stratosphere" (section 4.6). The presence of KAO investigators at Ames was beneficial for the program staff and for investigators from other institutions.

When the KAO began science operations in 1974, the scientists in the Astrophysics Branch of course began observing from it, as well as providing support for it. They and other astronomers in the Ames Space Science Division—about a dozen in all—competed successfully for about 20 percent of the total number of astronomy flights over the program's lifetime, via the same peer review process experienced by all investigators.

The strong in-house observing and instrument-development programs were part of a larger, relevant science community at and near Ames. Visiting scientists benefited from the presence of the other Ames scientists and detector development engineers, and from the proximity of local universities (seven within a 30-mile radius) where related research was ongoing. These contacts produced synergy in the research activities, via informal discussions over lunch, ad-hoc meetings, seminars, etc. For example, meetings of the Center for Star Formation, a NASA-sponsored program jointly hosted by U.C. Berkeley, U.C. Santa Cruz, and Ames, were often attended by researchers visiting Ames in conjunction with their KAO observations. The local science environment was certainly a contributing factor in the effectiveness of the KAO.

4.9 ANNUAL OPERATING COSTS AND PRODUCTIVITY METRICS

The cost of the KAO operation in 1995, not including the cost of civil service (CS) labor, is not known. However, the current (2013) CS labor rates are known. This permits an estimate of what it would cost to operate the KAO now.

For FY1995, NASA Headquarters provided $13M to operate the KAO. Of this funding, $3M was for grants to participating science teams. The remaining $10M was for fuel and "other costs," which included contractors, maintenance, etc. Not included in the $13M was the cost of the KAO 32-person CS operations staff and CS engineering/shop support (see section 4.7.1). Extrapolated cost estimates of the program elements are as follows:

Science grants: The inflation between 1995 and September 2013 is a factor of 1.53, based on the consumer price index (CPI, ref. 27). Scaling for inflation, the $3M FY1995 expense for science grants would become $4.6M in FY2013.

Fuel costs: The C-141 fuel burn rate was about 2,000 gallons per hour for typical 7.5-hour flights. In 1995, jet fuel (JP-8) cost about $1.20 per gallon. Hence for the average 635 flight hours annually (section 4.4.1), the fuel cost in 1995 was about $1.5M, leaving $8.5M of "other costs" provided by NASA Headquarters. The current (September 2013) fuel price is $4.26 per gallon (ref. 28), so the fuel cost now would be $5.4M per year.

"Other costs": Applying the CPI factor of 1.53 to the 1995 amount of $8.5M, the "other costs" become $13.0M for 2013.

Civil service labor: The average non-reimbursable cost (including salaries, benefits, direct management overhead, and default travel allowance) of CS labor at NASA Ames for FY2014 is about $0.185M/FTE (ref. 29). Thus, in 2013 dollars, the cost of the 32 FTE CS labor would be $5.9M.

Total 2013 cost: Adding these elements, the current annual KAO operating cost would be $28.9M. This number can be used to evaluate some operational metrics.

One cost-effectiveness metric is the cost per observing hour. There were an average of 518 research flight hours accumulated on the average of 72 flights per year. Ascent plus descent took a total of about an hour per flight, so the number of successful flight hours (SFH) averaged about 446 per year. The SFH are hours when the observatory is actually making observations, that is, hours when it is at operating altitude with the telescope and a science instrument capable of taking useful data. The current cost per SFH then would be about $65,000, or close to $1,000 per minute.

A similar metric is the number of SFH per equivalent full-time (FTE) person supporting the project. This turns out be 7.2 hours per FTE, which is roughly one flight per person. The numbers are summarized in table 9.

Table 9. Current Epoch (2013) Annual KAO Operations Costs/Productivity

Item:	Science	CS Labor	Fuel	Other	Total	SFH	Cost/SFH	FTE	SFH/FTE
Value:	$4.6M	$5.9M	$5.4M	$13.0M	$28.9M	446	$65k	62	7.2
%:	16	20	18	45	100				

4.10 KAO LIFETIME RECORD, COSTS, AND SCIENCE METRICS

Table 10 summarizes some significant aspects of the science program over its lifetime (1974–1995). A far more thorough description through 1990 is given by Larson in reference 2.

Table 10. KAO Lifetime Science Program Summary

Instrument Teams	Instruments Flown	Instruments Flown/Year	Principal Investigators	Total Investigations	Investigations per Year	Total Publications
33	>50	~10	126	~510	~25	>1000

A 1993 report prepared for NASA Headquarters (ref. 30) compared cost effectiveness of past and then current NASA Astrophysics Missions, including the KAO, on the basis of several criteria. The data include costs for Development, Mission Operations, and Data Analysis. The latter is the amount granted to the U.S. science community for participation in the science program including grants to NASA CS scientists. In table 11, cost values from the report for the KAO have been updated to include operations through 1995, CS labor for operations and upgrades, and inflation through FY2013.

Table 11. KAO Lifetime Costs in 2013 Dollars				
Category:	Development	Operations	Data Analysis	Total
Full Cost:	$110M	$380M	$107M	$597M
%:	18	64	18	100

The 1993 report compared the KAO with contemporaneous NASA Astrophysics Missions on the basis of two potential metrics for judging their cost effectiveness; these are items 1 and 2 in table 12. Item 1 is the fraction of cost related directly to the science output; the KAO exceeded all other astrophysics (space) missions. The second item is a measure of the cost of high-public-appeal science findings, for which the KAO ranked third lowest (third best). The other metrics in table 12 are additional criteria for program effectiveness.

Table 12. Some Science Mission Metrics for the KAO (2013 $)	
1. Support for science research: data analysis cost / (development + operations costs)	20%
Highest percentage of all NASA Astrophysics Missions up to 1993	
2. Science prominence: cost per *Science News* annual "most important stories" citation	$159M
Third lowest among NASA Astrophysics Missions up to 1993	
3. Cost (including facility development) per publication	~$600k
4. Cost (including facility development) per science grant	~$1.2M
5. Oversubscription factor for observing time (circa 1994)	~2.5
6. Oversubscription factor for proposed investigations (circa 1994)	~1.5
Cost numbers include estimates for civil service labor	

In the Introduction, success was defined as "safely achieving a science productivity that approaches the maximum possible for available resources." Based on that criterion and the KAO performance data presented here, the KAO program was indeed a successful scientific endeavor. The fact that the oversubscription rates were not higher may be due in part to the "closed shop" perception among some members of the science community (section 4.8.2), to the fact that infrared astronomy was still a somewhat unfamiliar discipline in the science community, and to the lack of concerted program promotion efforts commonly seen for astronomical facilities at science meetings today.

The scientific output of a mission can almost always be increased by increasing the resources expended on it. At some point, however, as the resources are increased, the *rate* of increase of scientific productivity will begin to decline. That is considered by many managers to be the optimum funding level for the mission.

Despite the success of the KAO and appeals by associated scientists and staff to increase its operating budget to permit a higher flight rate, that never happened—the flight rate remained substantially constant (fig. 9). However, the budget *was* increased about 50 percent (in constant-year dollars) over the program's lifetime, largely to compensate for increasing fuel costs and more frequent deployments (table 4). Certainly productivity of the KAO could have been increased with additional funding, and its proponents believed it was not operating near the potential peak of its productivity / cost curve.

In addition to its scientific merits, the KAO made three other major contributions to the discipline of astronomy: (1) considerable new technology for science instruments was developed and demonstrated, (2) numerous instrumentalists—many of them young—gained valuable experience with infrared technologies applicable to space astronomy, and (3) the cadre of some 600 participants in the KAO science program, and in particular the 126 Principal Investigators listed in table B1, became a major asset to the field. The latter claim is substantiated in the following section.

4.11 THE VALUE OF AIRBORNE ASTRONOMY TO THE ASTRONOMICAL COMMUNITY

Beyond its direct contributions to astronomy, the value of the KAO program to the community is manifested in the subsequent contributions of its alumni. This value is evinced in part by the recognitions received by its participants. Some of the awards earned by astronomers experienced with airborne astronomical instrumentation are listed in table 13. We apologize to the deserving individuals not listed here. Nine of the 17 awardees were airborne instrument team leaders. These awards, while not in every case related directly to research done in the airborne program, demonstrate (1) its appeal to talented individuals for creative application of advanced technologies, and (2) its excellent opportunities for mentoring and developing researchers' skills in observational astronomy and instrumentation. That 5 out of 11 of the American Astronomical Society Weber Awards for instrumentation have gone to researchers with airborne astronomy experience attests not only to the effectiveness of the program in fostering opportunities for new instrumentation developments by individual teams, but also to the potential for rapidly advancing infrared and submillimeter technologies.

Many of the 126 KAO Principal Investigators listed in table B1, especially the instrument team members, have gone on to play major—often leadership—roles in ground-based and spaceborne astronomy programs. Some 40 of these are listed in table B2, along with some of their subsequent accomplishments. The data in tables 13 and B2 are clear evidence for the outstanding record of the KAO in attracting and developing productive talent for the astronomical community.

Table 13. Some Awards Received by Astronomers With Airborne Experience	
AAS Pierce Prize for outstanding achievement in observational astronomy over the past 5 years for researchers under 36 years old	Eric E. Becklin,# Doyal A. Harper,*# Reinhard Genzel,# Harriet L. Dinerstein, Kristen Sellgren*
AAS Cannon Award for outstanding research and promise for future research by a woman within 5 years of receiving her Ph.D.	Harriet L. Dinerstein, Suzanne Madden
AAS Weber Award for Astronomical Instrumentation leading to advances in astronomy	Frank J. Low,# Thomas G. Phillips,# Harvey Moseley,*# James R. Houck,# Thijs de Graauw#
ASP Bruce Gold Medal for a lifetime of outstanding research in astronomy	Martin Harwit,# Frank J. Low#
ASP Muhlmann Award for innovative advances in astronomical instrumentation	John H. Lacy, Michael Skrutskie
Nobel Prize for fundamental work in quantum electronics	Charles H. Townes#
MacArthur Foundation Award for astrophysics	John E. Carlstrom
Pawsey Medal (AU) for excellence in experimental physics	John W. V. Storey

AAS: American Astronomical Society
ASP: Astronomical Society of the Pacific
* indicates Ph.D. thesis included data from airborne observations
# indicates team leader for development of airborne astronomy instrument(s)

4.12 FACTORS FOR SUCCESS—WHY IT WORKED SO WELL

The preceding description of the KAO program, reasons and evidence for its success, and possible improvements are based on the KAO archives, on opinions of associated scientists (both at Ames and elsewhere) and staff members, and the authors' experience. Some accounts were recorded, e.g., those of Connors (ref. 31), and of Rentch and Zaitzeff (ref. 32). From these sources, table 14 lists many of the factors, major and minor, that contributed to the successful operation of the KAO or that may have enhanced its productivity if done differently.

The items in table 14 are certainly not all of equal importance, although all are features that contributed to program productivity. Rationale for them has been discussed above.

In retrospect, among all the KAO factors for success, one that stands out is the efficiency and synergy of co-located facilities and personnel. This includes not only the KAO mission and flight operations, but also the immediate access to the science labs, machine shops, test facilities, and related engineering, technical, and science support personnel at NASA Ames. Corroborating this view was a 1995 letter to NASA management from the chairman of the KAO Users' Group, Dr. Gordon Stacey, and endorsed by over 100 astronomers, including a majority of all the KAO Principal Investigators listed in table B1. The letter recommended consolidated SOFIA operations at NASA Ames, as had been the scenario for the KAO (ref. 33).

Table 14a. Airborne Astronomy Factors for Success: Science, Operations, and Upgrades

	KAO	*Benefit
Science:		
Instrument team support: for Guest Investigators' observations	Yes	1,2
Investigators: encouraged to participate in observations	Yes	2
Instrument development: supported for innovative concepts and new technologies	Yes	4
Make-up flights: granted to investigators for unforeseen facility downtime	Yes	3
Facility instruments/service observing: supported by the observatory	No#	1
Data archiving: from *all* science instruments in calibrated, useful formats	No#	1
Guest Investigators: allowed to submit proposals independent of instrument teams	Rarely#	1
Operations:		
Co-location: proximate science, mission, and flight-operations personnel and facilities	Yes	3
Staff attitude: focused on getting data, not perfect functioning of all systems	Yes	3
Matrixed staff: for ground servicing and in-flight operation of mission systems	Yes	3
Flight-planning: software tools available for users to optimize observing options	Yes	1, 3
Access to telescope cavity: required for science-instrument installation	Yes#	3,4
On-board communications: multiple, real-time video displays and mission consoles configured for nonverbal information transfer	Yes	2,3
In-flight contingencies: on-board science and mission personnel qualified to meet	Yes	2
Star-field recognition: automated with facility video cameras	No#	3
Access to science instruments: while operating in flight	Yes	2,3
On-board bunks for crew rest: available in flight	Yes	1
On-ground communications: daily briefings, "squawk" board	Yes	3
Upgrades and Maintenance:		
Continuous improvement: to enhance cost-effective performance	Yes	2,3
Hardware and software: commercial products extensively used where practical	Yes	3,4
Software: minimal approvals needed for mission systems and science instruments	Yes	3,4
Airworthiness approvals for instruments: simple, straightforward, efficient	Yes	1,3
Aircraft maintenance: reliable, convenient, economical	Yes	3

* Benefit Categories: 1. Science productivity and community participation 2. Data acquisition efficiency 3. Facility operational efficiency 4. Science potential enhancement
# Feature that may have improved the KAO productivity if changed.

Table 14b. Airborne Astronomy Factors for Success: Facilities		
Hangar, Incorporating:	**KAO**	***Benefit**
Safety systems: NASA-approved for aircraft operation	Yes	3
Laboratories: for science-instrument integration and maintenance	Only 1#	3
Staff offices and work areas: for scientists, mission staff, and aircraft crews	Some#	3
Machine shop: small, accessible to participants	No#	3
Aircraft systems storage: for small parts, large equipment	Limited#	3
Mirror coating facility: vacuum tank, pumps, crane, etc.	No#	3
Airfield:		
Runway: low-use, with high take-off priority for the observatory	Yes	3
Airspace: minimal proximity to restricted areas and international borders	Yes	3
Local weather: moderate	Yes	3
High-altitude weather: low tropopause	~6 Months	
Security: convenient access provisions, especially for foreign nationals	Sporadic#	1,3
On-Site Facilities and Staff:		
Fabrication shops: well-equipped machine, sheet metal, electronics, instruments	Yes	3,5
Engineering staff: mechanical, electronics, aeronautical	Yes	3,5
Relevant science research groups: airborne observers, theorists	Yes	1,4
Astronomical infrared detector laboratory: array evaluation expertise	Yes	3,4
Local Infrastructure:		
International airports: within 30 miles	3	1
Commercial technical parts' suppliers: equipment maintenance, repair, upgrades	Yes	3,5
Lodging and meal service: convenient for visiting science teams and air crews	Yes	1,3
Universities with relevant science programs: within 30 miles	7	1

* Benefit Categories: 1. Science productivity and community participation 2. Data acquisition efficiency 3. Facility operational efficiency 4. Science potential enhancement 5. Facility upgrade capability
# Feature that may have improved the KAO operation if changed.

Similarly, a 1996 independent analysis of airborne science operations by the NASA Inspector General (IG) concluded: "In general, the costs associated with the operation of these [science research] aircraft are principally driven by research support requirements, i.e., modifications, research hardware, flight hours, and travel as required to support the technical program" (ref. 34). Because flight operations necessarily occur at the operations site, this conclusion clearly supports the close physical association of all program elements. The basic finding in the IG report is confirmed by the manpower distribution for the KAO (table 6)—the staff dedicated to mission and science operations was over twice that for aircraft operations. The value of a tightly knit operation is further substantiated by the excellent efficiency of operations during KAO deployments when the mission staff, aircraft crew, and science teams were billeted in close proximity to the airfield where the observatory was based, e.g., in Christchurch, New Zealand.

5. SOFIA

5.1 JUSTIFICATION

In the late 1960s it was recognized that a telescope that would operate in the stratosphere and be at least three times the diameter of the KAO's 91-cm optic would be both cost effective and far superior scientifically. This rationale was strengthened as SOFIA's anticipated performance and design evolved, as is discussed below.

SOFIA's larger aperture, increased observing time, and more powerful focal plane science instruments will result in tremendously improved capability over the KAO. In terms of cost-effectiveness, consider for example, the integration time $\boldsymbol{t}$ required to obtain the same signal-to-noise ratio using two telescopes whose diameters differ by a factor of $\mathcal{R} > 1$. Many of the observations done from the KAO and SOFIA were (will be) at infrared wavelengths where the images of compact sources are diffraction limited by the telescope, and the detectors are background-noise limited. Under these conditions, applicable to seven of SOFIA's nine originally approved first-generation instruments, $\boldsymbol{t}$ is proportional to $\mathcal{R}^{-4}$ (ref. 35). For the KAO and SOFIA, $\mathcal{R} = 3$, so for this common class of observations, SOFIA data acquisition rates will be about 80 times faster per detector to achieve the same signal-to-noise ratio on the same compact source.

The angular resolution of a telescope with diffraction-limited optics is roughly λ / D (radians), where λ is the wavelength and D is the telescope diameter. At wavelengths λ greater than about 5 µm where most SOFIA observations will be made, SOFIA will produce three times higher angular-resolution images than the KAO or Spitzer (ref. 36 and fig. 32).

Figure 32. Comparison of KAO and SOFIA beams (λ / D) on Saturn. The three times smaller diffraction-limited beams for $\lambda > \sim 5$ µm of SOFIA will allow resolution of features on astronomical objects not possible with the KAO.

Further, with about nine times the collecting area, SOFIA will measure much fainter objects than was possible with the KAO.

These advantages ensure a much higher data return per SFH (or per dollar spent) and often significantly higher data quality for SOFIA than its pioneering predecessor could produce.

Of course cryogenic telescopes in space have a vast sensitivity advantage for some important types of instruments because of the reduced infrared background (noise) from the telescope. However, the basic advantages of the KAO will be retained by SOFIA. Its long lifetime and easy accessibility will ensure that the benefits of improving technology can continue to be exploited. For example, its instruments can incorporate larger format, more sensitive array detectors, can include a number and variety of instruments to cover a wider range of wavelengths and spectral resolutions, and can be less expensive because they do not need to be remotely operated and can be readily serviced. Finally, as with the KAO, SOFIA's programmable mobility will provide access to the entire sky and to ephemeral events.

5.2 HISTORY

SOFIA owes its existence to the airborne science program at NASA Ames and, in particular, to the cohorts of its flagship facility—the KAO. This section contains a review of some of the individual contributions and events, which occurred primarily during the lifetime of the KAO, that led to the successful promotion of, specifications for, and design of SOFIA. The authors apologize in advance to the many whose efforts are not cited in this short review.

5.2.1 1970–1979

Starting in 1970, at the beginning of every decade the U.S. National Academy of Science sponsors an "Astronomy and Astrophysics Decadal Survey" report, which is organized by the National Research Council and written by an ad hoc Astronomy Survey Committee. The study, named after its chairperson, is used to guide NASA and the National Science Foundation in their prioritization of projects for study and/or development, and for research to be encouraged in the following 10 years.

The development of the KAO, having been successfully promoted within NASA by Ames management, was well underway (section 4.1) by the time studies for the 1970 Decadal Survey ("Greenstein") report (ref. 37) were ongoing in the late 1960s. The Infrared Panel of the Survey Committee included (among others), Professor Frank Low who had pioneered far-infrared observations from the Ames Learjet, and Drs. Eric Becklin, Jim Houck, and Harold Larson who would become KAO users. No doubt it was largely their contributions that led the Survey Committee to recognize not only results from the Learjet and the expected value of the anticipated KAO with its 91-cm telescope, but also the need for a larger facility. The final report recommended that NASA initiate studies for a 3-meter class, stratospheric telescope for infrared astronomy.

This recommendation in the 1970 Decadal Survey Report encouraged predisposed managers and scientists at Ames to pursue considerations for a Large Airborne Telescope (LAT). A year after KAO operations began in 1974, Ames Director Dr. Hans Mark circulated a letter to members of the astronomical community promoting the LAT idea, but the suggestion was not broadly supported.

A 1977 study by Boeing, requested by KAO managers Bob Cameron and Carl Gillespie, and supported by Ames engineer Ted Brown, featured a 2.5-meter KAO-configuration telescope installed aft of the wing in a Boeing 747. The 747 was a logical choice for the platform, based on its fuselage diameter and flight profile. This study was not widely circulated.

5.2.2 1980–1984

In preparation for the 1980 Decadal Survey (field) report, many in the infrared community opted to promote the concept of a 10-meter class, Large Deployable Reflector (LDR) in space to operate in the far-infrared/submillimeter part of the spectrum; its construction was then recommended in the report (ref. 38). Resulting substantial studies funded by NASA in the 1980s showed the LDR concept to be infeasible with available or anticipated technology. The Shuttle Infrared Telescope Facility (SIRTF) narrowly missed being recommended for development in the report. A LAT was not promoted for inclusion as a recommendation. However, in 1980 the concept was presented at the first meeting of the International Astronomical Union devoted to infrared astronomy by the KAO Facility Scientist Dr. Ed Erickson. The potential for a LAT was amply supported by the results from the first 6 years of KAO operations.

In 1982, Dr. Peter Mezger, a senior professor from the University of Bonn, Germany, toured the KAO at Ames. He and Erickson discussed the lack of NASA funding needed to augment the KAO flight rate, and incipient thinking at Ames about a larger KAO successor. Mezger inquired about the possibility of German participation, and was referred to Dr. Nancy Boggess, the infrared programs manager at NASA Headquarters. German interest was also no doubt stimulated by Dr. Reinhard Genzel, a German researcher doing KAO observations with Professor Charles Townes' group from the University of California at Berkeley.

A 1982 workshop on KAO aero-optical effects, organized by Erickson, included five experts from around the U.S. One, Dr. George Sutton, continued his interest in the SOFIA problem, eventually publishing the most complete predictions for the optical disturbance ("seeing") as a function of wavelength in 1998 (ref. 39). Dr. Ted Dunham, Professor Jim Elliot, and aerodynamicist Dr. Bill Rose carried out measurements of seeing and aerodynamic parameters on KAO engineering flights from 1983 through 1989, exploring the effects of temperature, altitude, and Mach number (ref. 25). These measurements became the basis for specifications of SOFIA's thermal cavity environment.

In 1984 an Airborne Astronomy Symposium (ref. 3), the brainchild of KAO user Dr. Harley Thronson, was held at Ames, with 48 papers describing results from the first decade of KAO observations and related topics. Among the over 100 participants were Genzel, Mezger, and Mezger's protégé, Dr. Hans-Peter Roeser, who had been observing on the KAO with his team from Bonn. Professor Roger Hildebrand presented a short paper entitled, "The Large Airborne Telescope." In this time frame also, Professor Martin Harwit produced the first market-worthy color brochure extolling the KAO accomplishments and potential of airborne astronomy. In the years that followed, both Hildebrand and Harwit continued to promote the LAT concept in the science community and at NASA Headquarters, as did other KAO users.

5.2.3 1985–1989

Early in 1985, KAO science coordinator Carl Gillespie (table 7) coined the name SOFIA. Ames scientists and managers began presenting descriptions of SOFIA to a variety of NASA review panels and Headquarters' officials, consistently winning endorsements, a process that was continually intensified until the project was eventually approved for development.

That year, with funding taken from the KAO program, the SOFIA Study Office was established at Ames, with Gary Thorley as manager, Nans Kunz as principal engineer, and Erickson as lead scientist. Coordinated in-house efforts at Ames were established between the Study Office, the Engineering Directorate led by Bob Eddy, and scientists in the Astrophysics Branch led by Dr. Larry Caroff who had assisted with observations on the Learjet and KAO.

Discussions were begun with Boeing and with the German Bundesministerium für Forschung und Technologie (BMFT), which subsequently morphed into the Deutsche Forschungs- und Versuchsanstalt für Luft- und Raumfahrt (DFVLR), then Deutsche Agentur für Raumfahrtangelegenheiten (DARA), and thence into DLR, the German Aerospace Center. It was agreed that the U.S. would supply the aircraft and mission systems, and Germany would provide the telescope and about 20 percent of support for operations in return for a comparable amount of observing time for German astronomers.

Also in 1985, what would become the SOFIA Science Working Group (SSWG), including participation by German scientists, was established as an offshoot of the KAO Users Subgroup. For the remainder of the KAO's lifetime and beyond, the SSWG provided science perspective to NASA and DLR, advised study teams on technical issues and requirements, ensured understanding between U.S. and German science communities, and advocated the project with colleagues, advisory groups, the public, and elected officials.

In 1986, NASA funding for the SOFIA Study Office began ramping up. Erickson and Professor Paul Harvey organized a SOFIA Technology Workshop at Ames. Twenty invited papers were given on topics including concepts for the telescope configuration, optical designs, lightweight mirrors, aircraft (only Boeing 747) modifications and performance, air stream control and effects, telescope stabilization, etc. (fig. 33). All consideration was given to installing the telescope ahead of the wing to permit the largest possible telescope. Among seven German participants were Alfred Dahl representing the DFVLR, Dieter Muser of Maschinenfabrik Augsburg-Nürnberg (MAN GHH), and the scientists Genzel and Roeser. The latter maintained indirect involvement the German effort, both with the telescope studies and subsequently with its development.

Later in 1986, substantial U.S.-German collaborative efforts were firmly established, both at the working level and between upper-level management of DFVLR and NASA Headquarters.

Figure 33. 1986 artist's concept of early SOFIA design. The telescope diameter depicted is 3.5 meters.

Initial industrial studies were begun both in Germany and in the US. NASA contracted with Boeing Military Airplane Co. (BMAC) in Wichita, Kansas, to investigate the major aircraft modifications required. DFVLR let contracts (managed by Alois Himmes) with two competing teams in Germany, MAN & Zeiss on one team, with Dornier & Zeiss on the other, to explore the feasibility of developing a larger KAO-like airborne telescope. With these parallel efforts began the establishment of interface requirements between the U.S. and German systems.

By 1987, efforts managed by the SOFIA Study Office had expanded to include about 25 people from different Ames organizations. Sufficient technical studies had been completed to formulate a credible plan for development. A NASA Non-Advocate Review vetted the plan.

Also in 1987, Dr. Jackie Davidson, whose Ph.D. thesis was based on observations she made from the KAO, joined the efforts at Ames to develop science rationale and technical specifications for SOFIA. Anticipating a possible NASA budget initiative, some KAO users—usually armed with explanatory materials prepared by Davidson and Erickson and the SOFIA Study Office—began contacting relevant congressional offices to apprise them of the concept.

A congressional information program for SOFIA was strategized and coordinated by Drs. Dan Lester and Harley Thronson, both of whom had earned their Ph.D.s using data they had obtained observing from the KAO. This unusual (for a program not in the NASA budget request) early effort was continued and expanded until SOFIA was approved for development.

Early in 1988, NASA Administrator Dr. James Fletcher's appeal for development funding in FY1989 was finally rejected by the federal Office of Management and Budget (OMB) because of fiscal constraints at NASA. Intensive collaborative efforts with DLR and German scientists continued in anticipation of a new start the following year, but by mid-1988 it became clear that NASA's limited Space Science budget would not enable an FY1990 start for SOFIA.

The BMAC cost estimate for aircraft modifications needed to install the telescope ahead of the wing were higher than expected. Although a Boeing engineer suggested that installing the telescope aft of the wing would be simpler and cheaper—a concept that had also occurred to the Ames engineering team—the baseline plan retained the telescope installation ahead of the wing.

A serious technical concern anticipated for SOFIA was the management of aerodynamic, aero-acoustic, aero-mechanical and aero-optic (seeing) effects of the airflow over and into the large open-port cavity housing the telescope. Relevant passive-control candidates were a porous upstream fence (used on the Learjet and KAO) and a contoured aft ramp, the latter having been strongly advocated by Ames aerodynamicist Donald Buell as early as 1976 in connection with the KAO. To examine the SOFIA seeing issues, a workshop was held at Ames in 1989. Kunz and aero-engineer Bill Rose initiated plans to carry out wind tunnel tests at Ames.

Also in 1989, as part of project planning, Professor Hal Larson drafted a Memorandum of Understanding between the U.S. and Germany, which would see many iterations and years before approval by the two governments. A second Non-Advocate Review approved the Ames plan for SOFIA development, but again NASA's budget was deemed inadequate.

KAO users had begun vigorously campaigning for an endorsement of SOFIA in the 1990 Decadal Survey report well before the committee was formally formed (with John Bahcall as chair) in 1989. This effort continued until the report was completed. Later that year, the 1977

Boeing report featuring the telescope installed aft of the wing mysteriously surfaced and was distributed to the Study Office.

5.2.4 1990–1995

Soon after the fall of the Berlin Wall in November 1989, it became apparent that financial repercussions could limit or delay German involvement in SOFIA. NASA Associate Administrator for Space Science, Dr. Lennard Fisk said, "We will build SOFIA, but not without the Germans." A third Non-Advocate Review in mid-1990 approved a plan that included German participation, to begin SOFIA development in FY1991. As expected, however, the DFVLR was forced to defer its formal commitment to SOFIA, although Mezger and others continued to promote German participation.

Undaunted, the SSWG continued vigorous promotion of the SOFIA concept (as did their counterparts in Germany), as inputs to the 1990 Decadal Survey (Bahcall) Report were being finalized. Cooperative efforts between SOFIA and SIRTF scientists resulted in the report (published in 1991) recommending construction of both facilities during the 1990s, in recognition of their complementary capabilities and the potential of infrared observations not possible from ground-based sites (ref. 40). John Bahcall remained a staunch advocate for both projects (and other activities recommended in the report) throughout the decade.

Armed with the Bahcall report recommendation, the SSWG expanded its membership and efforts to promote SOFIA at NASA Headquarters and in Congress. Principle Headquarters' advocates were Caroff, who had taken a temporary position there in 1988 to help promote the project, and Mike Kaplan. Guidance in planning was provided by Charlie Pellerin, Director of the Astrophysics Division, and by his successor, Dr. Dan Weedman. In 1993, Professors Charles Townes and Jim Houck, and Erickson met with NASA Administrator Dan Goldin to apprise him of the merits of, and plans for, SOFIA.

Technical study efforts continued both in Germany and the U.S. (fig. 34). In 1990, tests with a 7-percent scale model of SOFIA that included a telescope cavity ahead of the wing, were carried out in the Ames 14-foot wind tunnel. After multiple iterations, effective management of the airflow was achieved with a curved aft-ramp. Following these tests, Rose tested an aft ramp configured for the KAO in the wind tunnel. Based on this work, an aft ramp was installed and flown on the KAO in 1992 (fig. C85) a project led by Ames engineer Paul Fusco. This configuration eliminated noise and airframe vibration previously experienced on KAO during observing flights after the boundary-layer fence was raised as required for opening the cavity door.

In 1991, when the future of German support was uncertain, the SOFIA Study Office initiated examination of an "all U.S. effort" to build SOFIA. However, NASA mandated that cost would not rise, including development of the telescope. Competing telescope studies by Lockheed, Hughes, and KAMAN Aerospace Corporation were sponsored. As a subcontractor to the latter, Dr. Hans Kaercher, a German engineer with extensive experience in telescope design at MAN, developed concepts that his MAN team included in their 1996 successful proposal to build the telescope.

Figure 34. Ames SOFIA Study Office team in 1990. Not all are pictured, and not all worked full-time on SOFIA. The manager, Gary Thorley, is at far right. His successor, Chris Wiltsee, is in the back row center, in front of the tree trunk.

Seeking to reduce costs, Kunz led a 1991 "descope" study proving that aircraft modifications for an aft-telescope installation would indeed be cheaper. Ames engineer Rick Brewster's NASTRAN model demonstrated that with reasonable reinforcement of the fuselage, its original strength and stiffness could be retained despite the large hole aft of the wing needed for the telescope to view the sky. The aft-cavity configuration was finally adopted as the baseline design in 1993.

In the early 1990s, explicit cooperation with scientists advocating SIRTF was coordinated to ensure the perception in Congress that both facilities were needed. In particular, Dr. Mike Werner and Erickson prepared literature directly and quantitatively comparing the relative merits of SIRTF and SOFIA, emphasizing, for example, SIRTF's exquisite photometric sensitivity and SOFIA's ability to accommodate large, high-resolution spectrometers. With such materials, SSWG members and many other KAO users contacted their congressional representatives, with NASA scientists participating to supply details about the project as needed. Lester and Edna Devore coordinated the similar efforts of school teachers who had participated in educational programs on the KAO (see section 4.6). Interested industrial concerns, anticipating significant contracted work, also contributed significantly to congressional advocacy.

Figure 35. 1994 Airborne Astronomy Symposium announcement letterhead.

A second, far larger Airborne Astronomy Symposium was held at Ames in July 1994 (fig. 35). The proceedings (ref. 1) list nearly 250 participants and include 133 papers, concluding with a comprehensive rationale and concept for "SOFIA, the Next Generation Airborne Observatory" by Erickson and Davidson. Strong German interest was evident, and numerous, fruitful off-line discussions regarding SOFIA took place. That fall,

NASA's budget request for FY1996 included funding to start SOFIA development. The request was approved by the OMB and included in the President's proposed FY1996 budget in early 1995.

That summer of 1995, Congress was deliberating this proposed budget. To encourage approval of SOFIA, a reception for congressional representatives and staffers was held at the German embassy in Washington D.C. The event was organized largely by Dr. Hans-Peter Roeser, who had observed from the KAO with his team from Bonn. Enthusiastic introductory talks were given by Drs. David Hollenbach (Ames) and Reinhard Genzel, and by NASA Chief Scientist, Dr. France Cordova. Many scientists from the SSWG participated, using the evening to pointedly promote the project with the 100 or so Capitol Hill people in attendance. Within a month, Congress approved a budget including funding for SOFIA development.

Top-level management at NASA Headquarters promptly imposed a new management plan for "privatization" of the project, requiring development to be managed by the eventual operations contractor. That fall (of 1995), the KAO was decommissioned as had been planned, to transfer its \$13M operating budget to the SOFIA development effort. At the time, it was expected that SOFIA would be flying in 2001.

Of course, by 1995 all known technical issues had been examined. These studies would evolve to result in preliminary designs for major SOFIA systems. For example, after the initial 1990 wind tunnel test, four subsequent tests were carried out at Ames with the telescope cavity aft of the wing. The last of these (in 1997) included an articulating model of the telescope and partial external cavity-door geometry on a Boeing 747SP (B747SP) fuselage (fig. 36). Rose and Kunz used the results of these tests to optimize the geometry of a contoured aft ramp, and to evaluate its aerodynamic effects and the air loads on the telescope. The resultant design was adopted for use on SOFIA.

Figure 36. SOFIA model in the Ames 14-foot wind tunnel (1997).

Vastly numerous other topics were examined, many capitalizing on experience with the KAO. To name a few: (a) determination of the elevation range for SOFIA to be 20–60 degrees to enable viewing more of the southern sky when flying over the U.S. than was possible from the KAO (range 35–70 degrees); (b) measurement of B747SP vibrations and inherent attitude stability in flight; (c) preliminary design for the large cavity door; (d) calculation of the light from jet-engine exhaust scattered into the focal plane by Ann Dinger, based on measurements by Ted Dunham of the exhaust plumes on NASA's B747 Shuttle Transport Aircraft; (e) measurement of the boundary layer on the same plane by Rose; (f) NASTRAN analyses of airframe loads, and computational fluid dynamics analyses of the altered airflow due to the presence of the large hole in the fuselage; (g) requirements definition for the aircraft garage, resulting in plans to modify Ames hangar N-211 where SOFIA was to be housed; (h) prediction of flight profiles based on known B747SP characteristics, and estimated telescope and mission system weights; (i) a spherical rotation-isolation system for the telescope, incorporating a large spherical bearing; (j) pointing stability predictions based on anticipated telescope structure and air-loads measured in the wind tunnel tests, coordinated by Kunz and modeled by Controls Engineer Paul Keas; (k) an analysis of turbulence events to be expected by Jackie Davidson; (l) a myriad of reports documenting system requirements, such as the Science Requirements Document by Erickson and Allan Meyer, the Telescope Requirements Document assembled by Jerry Hirata, and the Aircraft Modification Requirements Document organized by Kunz; (m) formulation of a comprehensive Work Breakdown Structure for managing the development; and (n) plans for managing development with the Ames SOFIA Project Office acting as prime contractor.

Contracts for DLR and NASA to begin SOFIA development were finally signed in December 1996.

It is clear that the experience gained and enthusiasm generated during the 21-year operation of the KAO was the *sine qua non* for the successful promotion of, and planning for, SOFIA.

5.3 THE FACILITY

SOFIA has been developed and is now (December 2013) operated in a collaboration between NASA and DLR, the German Aerospace Center. Resources in the partnership are supplied and observing time allocated in a nominal 80/20 (U.S./German) ratio. The facility is a Boeing 747SP aircraft, provided and modified by NASA, carrying a 2.7-m telescope supplied and developed in conjunction with NASA by the DLR. SOFIA is designed to provide 960 successful flight hours (SFH) for research per year throughout a 20-year design lifetime. This duty cycle and a number of scientific instruments are expected to enable SOFIA to support on the order of 50 investigator teams per year. Figures 37, 38, and 39 depict important features of this beautiful observatory.

Figure 37. SOFIA on a test flight in 2007 near Waco, Texas, where the aircraft was modified. A new full-pressure bulkhead just behind the wing leaves the aft end of the fuselage unpressurized to accommodate the telescope. A two-segment door closes the telescope cavity except during observations.

Figure 38. Optics in the SOFIA telescope. With a diameter three times larger than that of the KAO telescope, SOFIA is a much more powerful tool for astronomy. The telescope was provided by NASA's partner, the German Aerospace Center (DLR).

Figure 39. First door-open test flight of SOFIA, 18 December 2009, near Palmdale, California. In this image the telescope's primary mirror has a protective cover. Test flights verified the effectiveness of the Ames-designed aft-ramp airflow control, as well as the large cavity-door assembly that was built there.

5.4 VISION FOR SOFIA

Users of the KAO provided extensive advice for the operation of SOFIA. For example, Dr. Gordon Stacey, then chairman of the KAO Users' Group, sent a letter to NASA management describing recommendations for SOFIA operations (ref. 33). The letter was endorsed by over 100 astronomers, including a majority of all the KAO Principal Investigators listed in table B1. In addition, the authors' original science requirements document for the project, SOF 1009 (ref. 41), incorporated a wide range of scientific, technical, and operational expectations for SOFIA, based on inputs from astronomers and engineers familiar with the KAO.

5.4.1 Founders' Recommendations for Operations

The SOFIA Science Working Group (SSWG), comprised largely of scientists who had been active in the KAO program, advised on plans for and promoted the approval of SOFIA beginning in the late 1980s. The SSWG envisioned SOFIA as retaining the valuable features of the KAO program: science inaccessible from the ground, innovative new science instruments and technologies, training of young scientists, flexibility in operations and scheduling of observations to accommodate targets of opportunity, continuous improvement on all timescales, and an effective education and public outreach program. The deliberations of the SSWG members benefitted from their KAO experience, as well as from their otherwise diverse experiences as astronomers.

In 1995, the SSWG framed a vision statement outlining recommendations for science operations. The statement is quoted verbatim here:

GOALS AND PROVISIONS FOR SCIENCE OPERATIONS

Considering SOFIA's heritage and science potential, the SOFIA Science Working Group—in conjunction with members of the astronomical community at large—recommended the following goals and consequent provisions to maximize the productivity of the SOFIA Science Operations Program (SSOP). Not all were features of the KAO program.

GOALS:
A: Maximize scientific productivity and discovery potential, and
B: Promote educational opportunities and public outreach.

PROVISIONS:

(1) Scientific direction of the observatory by an astronomer who is responsible for all aspects of its operation, particularly its performance and productivity. This arrangement gives the best assurance that the observatory's resources will be allocated in a balance which achieves the goals of the SSOP.

(2) Frequent independent peer review of proposed science and technology to assure not only the value of ongoing research, but also routine opportunities for the entire astronomical community to propose new research. This implies that the SSOP will encourage guest, as well as principal investigators, and will enable straightforward access by scientists from all nations.

(3) An instrument complement that exploits the observatory's full potential. It is essential that the SSOP will provide resources for continuous, innovative instrument and technology development over the projected twenty year lifetime of SOFIA. The objective is to make available state-of-the-art focal plane instruments that achieve performance determined by the physical limitations of the observatory.

(4) Deployment to remote sites and rapid response to transient phenomena. This implies not only that the SSOP must include routine deployments for extended southern hemisphere observations, but also that it must provide for self-contained expeditions to non-routine destinations to observe astronomical events such as occultations or eclipses.

(5) A vigorous flight program and highly efficient operations to assure effective utilization of the observatory. These imply maximizing—within practical constraint —the number of flight hours per year and the efficiency of operations, for example, in exchanging focal plane instruments on the telescope, in flight planning, and in observing. These also imply that facility upgrades are implemented as needed to improve performance.

(6) Stimuli for prompt publication and archiving of results. These will promote rapid access to observational results and their significance by the science community.

(7) Flexibility in the program to include different types of investigations will permit, for example, approval of multiple-year proposals and key projects as appropriate to benefit from the continuity afforded by SOFIA's long lifetime. However, some speculative, high-risk investigations should also be included in the observing program, as should rapid adjustments of the schedule needed to accommodate observations of unexpected ephemeral events by qualified investigators.

(8) A variety of educational activities, including development of stimulating programs for K-12 teachers, innovative science programs for undergraduates, graduate-student doctoral research opportunities, research opportunities for young scientists, and exposure of infrared astronomy to the general public and the media. SOFIA's accessibility, mobility, and lifetime give it tremendous potential in this arena.

These features are considered to be the expectations of the science community for SOFIA as a unique, world-class astronomical observatory. The purpose of development is to permit these goals and provisions to be fully achieved, while minimizing the resources required.

Regarding item 1, the Science Working Group believed that an astronomer directing the observatory would be better able and more inclined to balance the resources between science, mission, and flight operations to maximize SOFIA's scientific productivity than was done with the KAO organization (fig. 24).

Item 6 was prompted by the desire to expand the value of observations made by making them more generally available to the science community (see section 4.8.3). Archiving all data from the instruments in calibrated and useful formats would clearly enhance scientific productivity. The SSWG realized that complicating factors in achieving this goal would be the expected evolving configurations, observing modes, and performances of the non-facility instruments, the politics associated with non-U.S. instruments, and the expense in maintaining the archive, which would be substantial.

5.4.2 Realization of SOFIA Operations

SOFIA flight operations are based in Palmdale, California, and science operations are based at Ames. Accordingly, some valuable features of the KAO operational model are not possible, in particular, the highly beneficial co-location of facilities and personnel (refs. 33 and 34). Others, such as vigorous support for new science instruments have yet to be implemented (ref. 42).

Of course it cannot be expected that SOFIA will operate exactly as did the KAO. However, indeed, a number of the cited KAO *factors for success*, founders' recommendations, and KAO operating procedures will be incorporated in SOFIA's operation. Two examples: facility science instruments with service observing for investigators will be standard features, and ground-based verification of systems functionality, begun with "first light" tests of the telescope in 2004, continue to be utilized effectively (fig. 40).

As of the completion of this document (December 2013), SOFIA management plans to declare achievement of Full Operational Capability (FOC) in the coming year.

Figure 40. SOFIA "line op" 2009. As with the KAO, much testing of the telescope system and science-instrument installations can be done during operations on the flight line ("line ops") with the aircraft on the ground, as in this image taken at Palmdale, California, where SOFIA flight operations are based.

6. THE LONELY DARK NIGHT SKY

A pamphlet entitled "The Lonely Dark Night Sky" and subtitled "A Collection of Songs, Poems, and Other Writings from the Kuiper Airborne Observatory" was prepared by Wendy Whiting Dolci for the combined retirement party of two KAO stalwarts, Carl Gillespie and Jim McClenahan, on June 8, 1998. The small document was dedicated to these two steadfast, (now deceased) principals whose professional lives focused on achieving success of the KAO throughout its 21-year operating lifetime.

The KAO operations team was a diverse but close-knit group. Their camaraderie was based on mutual respect, and their common dedication to the observatory and its science mission. The booklet contained staff members' impressions from the days of KAO operations. Especially on extended deployments in New Zealand, some found time to compose these vignettes reflecting boredom, humor, and/or nostalgia. Some of the contributions were written for the KAO farewell ceremony that took place at Ames on September 29, 1995 (fig. 41). The reader should be aware that these epitaphs for the KAO were written with the understanding that its decommissioning would contribute to its offspring—SOFIA—and with the expectation that SOFIA would be flying in a few years. Despite these positive feelings, most attendees that day harbored bittersweet sentiments, which are mirrored in some of these ruminations. Some of the pamphlet's entries are captured below, reminders of the affection many had for the activity.

Sky Fever

Jim Cockrell, KAO Electrical Engineer, September 29, 1995

I must up to the skies again, to the lonely, dark night sky,
And all I ask is a telescope, and a star to steer her by;
And turbulence, and compressors whine and the PFC shaking,
And a grey mist in the tracker field, 'fore the grey dawn's breaking.

I must up to the skies again, for the call of the stratosphere,
Is a wild call and a clear call that I shall always hear.
And all I ask is a high jet stream with really low water vapor,
And a bright source with some broad lines for a grad student's paper.

I must up to the skies again, in my preflight ritual,
With the aperture wide open, and the tanks filled with fuel.
And all I ask is some oxygen, half way twixt earth and heaven,
And a coffee cup, and a good flight lunch, in the aft galley oven.

I must up to the skies again, to the starry black night sky,
Where the chopper's pulse and the engines' drone are a Kuiperman's lullaby
And all I ask is a good flight plan to get me through the night,
And a quiet sleep and a sweet, sweet dream, at the end of the data flight.

Apologies to John Masefield, author or *Sea Fever*, one of his *Salt Water Poems and Ballads*.

New Zealand Blues

Tom Connors, KAO Mechanical Engineer, circa 1991

Christchurch at last and the plane's touchin' down
the Heavens opened up, photons danced all 'round
through a cold chain link fence
to a frost covered rental car
it's a half minute ride to the Travelodge bar.

Boredom's a pastime that one soon acquires
'til ya get to the preflight and you're not even tired
Kickin' your heels till the time comes around
check the tanks one more time
get this jet off the ground.

(chorus)
Slow down mate, we're Kuiperwomen and Kuipermen
We've frozen our bones in this high flyin' can
you ain't seen nothin' 'til ya been
on deployment in New Zealand
based at the Travelodge Inn.
E-mail, faxes and preflights just don't seem the same
frustration all around, tempers gettin' hard to tame
your feet stay soakin' wet
as the weather beats you down
constant thoughts of when you'll abandon this town.

The Arts Center's fun but only two days a week
new places to go you constantly seek
you think things couldn't be worse
any other place would be nice
thank your lucky stars you're not on the Ice.

Supernova Calibrator Expialidocious

Brian Wright, KAO Airborne Data and Management System (ADAMS) staff, circa 1989

Supernova Calibrator Expialidocious
Found the target field except
the cameras weren't in focus.
Tried to take some data
but the pointing was atrocious,
Supernova Calibrator Expialidocious

Kuiper Blues

Jim Cockrell, KAO Electrical Engineer

Woke up this evening, put on my shoes;
Went to work singin' these Kuiper Blues.
Never did think I could sink so low,
Doin' odd jobs on the KAO.

Slam the doors and kick the tires,
Sign the checklist, light the fires.
Too late to bail out so look out below,
Doin' odd jobs on the KAO.

Operator set them gyros to spin.
MD wearin' his maniacal grin.
It's a Howdy Doody, Punch 'n Judy show,
Doin' odd jobs on the KAO.

We're up all night and it ain't no fun,
Countin' these photons one by one.
Signals or noise them PI boys don't know,
Doin' odd jobs on the KAO.

'Scope stopped trackin' and into a stall;
Computers up and died from somethin' terminal.
With the fuses blowin' and nothin' to show,
Doin' odd jobs on the KAO.

This bad flight coffee gonna make me sick;
I get no joy from my joystick.
But there's real recompense when the data flow,
Doin' odd jobs on the KAO.

Doin' odd jobs on the KAO;
Nights so long and flights so slow.
It's even more weird than I first feared,
Doin' odd jobs on the KAO.

A Farewell to the KAO

Ed Erickson, KAO Facility Scientist, September 29, 1995

In 1797 the United States Navy commissioned the frigate Constitution, which distinguished itself in a number of battles, earning the nickname "Old Ironsides" because of the resistance of her oak planking to enemy cannon balls. In the war of 1812, she defeated 5 British men-of-war and sank 12 merchant ships, salvaging only their rum. By 1830 the ship was in need of extensive repairs, perhaps the equivalent of a D-check in modern aircraft parlance, and the Navy decided instead to decommission her. The decision was protested by many people, including Oliver Wendell Holmes, who later became a famous author and physician. His protest was in the form of a poem, which I have taken the liberty to adapt for this ceremony marking the retirement of the Kuiper Airborne Observatory (fig. 41). With deepest apologies to Oliver Wendell Homes, who I think would approve:

KAO

Aye, tear her gleaming ensign down!
Long has it flown on high,
And many an eye has danced to see
That symbol in the sky.
Far down, bound by gravity,
The earthlings watched it soar;
The meteor of the stratosphere
Shall sweep the clouds no more.

Her telescope, once floating instruments
Attuned to get the data,
When winds were blowing her off course,
While zooming through the strata,
No more shall feel discovery's glee,
Or know the angst of sighs:
The harpies of the earth shall pluck
The eagle of the skies.

O better that her shattered hulk
Should sink beneath the wave!
Her thunders shook the mighty heights;
As deep should be her grave.
But as the Phoenix rose again,
Her spirit will soar on:
To fly in the millennium,
When a new age will dawn.

PS: You can visit Old Ironsides in Boston Harbor. She is in first-class condition, maintained as a national monument—the oldest commissioned ship in the world.

Danny Boy was a song often sung by staff members together in informal, relaxing situations after a few libations and reminiscences had been shared. These occasions were most common on extended deployments when, distant from home, KAO supporters found more opportunities to bond in off-duty hours.

Danny Boy

Frederic Weatherly, 1910

Oh, Danny Boy
The pipes, the pipes are calling,
From glen to glen
And down the mountain side.

The summer's gone,
And all the flowers dying.
It's you must go, it's you must go
And I must bide.

But come ye back,
When summer's in the meadow,
Or when the valley's hushed
And white with snow.

It's I'll be there,
In sunshine or in shadow;
Oh, Danny Boy, oh Danny Boy
I love you so.

Figure 41. The KAO farewell and decommissioning ceremony 29 September 1995, in front of the KAO hangar N-248 at NASA Ames. Most of the KAO staff and many KAO investigators from around the U.S. attended. Involved astronomers and NASA managers agreed to terminate the KAO program to make its operating budget available for the development of SOFIA. The last KAO flight, supporting a brief education program called "Live from the Stratosphere," was flown the night of October 12–13, 1995.

7. SUMMARY

Early experience with airborne science, and in particular with astronomy at NASA Ames, led to the development and operation of the KAO. During its 21 years of operation, it logged nearly 13,000 research flight-hours, about 84 percent of all of its hours in the air. Of research flights scheduled, 88 percent were flown successfully, and only about 5 percent of attempted flights were aborted. On average, there were 25 observational programs per year, led by a total of 126 Principal Investigators over the lifetime of the program.

The KAO stimulated development of over 50 science instruments that produced data for over 1,000 publications, many containing revelatory scientific findings. Metrics for effectiveness of the KAO show it was highly successful relative to contemporaneous NASA Astrophysics Missions. Participants demonstrated technologies and gained experience that contributed significantly to subsequent NASA missions. Over 50 graduate students received Ph.D.s for airborne astronomy research. Numerous scientists who observed from the KAO have become leaders in the field of astronomy. Many members of the media and public in the U.S. and abroad visited the observatory, gaining a firsthand impression of this unique facility.

The KAO operation was characterized by a focus on the primary program goal to "get the data" while operating safely. For example, it featured science-instrument and mission-systems software requiring minimum review and approval; efficient airworthiness approval procedures for science instruments; flying, if possible, with just a minimum complement of mission systems operating; on-board, qualified mission personnel and scientists able to deal with in-flight contingencies; and incremental upgrades with minimal schedule impact.

The operation was lean and efficient, with a full-time equivalent staff of 26 civil servants and 31 support service contractors. The manpower distribution by tasks was 55 percent for mission operations, 28 percent for aircraft operation, 12 percent for management, and 4 percent for science support (table 6). Matrixed staffing was appreciable; for example, all the managers participated in flights as Mission Directors. The presence of the local science community and facilities enabled immediate response for relevant issues with the observatory and with visiting scientists and their instruments.

The KAO program described here led to a list of factors that contributed to its success, as well as some factors that may have enhanced its effectiveness had they been implemented. These *factors for success* (summarized in table 14), and the vision statement for SOFIA science operations (given in section 5.4.1) by the 1995 SOFIA Science Working Group, embody much of the wisdom accumulated during the KAO experience. Primary among these was the enthusiastic emphasis on scientific productivity that was *greatly* facilitated by co-location of the entire operation.

Success in all such activities is enabled extensively by a solid spirit of purpose, flexibility, teamwork, and objectivity in solving problems. These hallmarks of the KAO program formed the basis of the plans and great expectations for SOFIA. SOFIA's success—that is, safely maximizing scientific productivity with available resources—will be enhanced to the degree that its operation effectively harvests lessons rooted in its unique KAO heritage.

APPENDIX A. KAO FLIGHT STATISTICS 1972–1995

Summarized here are the annual flights and flight hours for the KAO program. Table A1 includes all flights. In all of the tables an attempt has been made to reconcile minor discrepancies in the archived statistics. Most, but not all, of the aborted flights were astronomy research missions, but no attempt was made to distinguish them from other categories here.

Table A1. Kuiper Airborne Observatory, Flight Operations Summary FY1972–FY1996

Flight Type	Aircraft Maintenance		Pilot Proficiency		Ferry		Engineering		Aborted		Other Research		Astronomy Research		Totals	
Fiscal Year	Flights	Hours	Flights	Hours	Flights	Hours	Flights	Hours	Flights	Hours	Flights	Hours	Flights	Hours	Flights	Hours
1972					2	6.0									2	6.0
1973	4	5.6	20	76.2											24	81.8
1974	20	40.8	9	30.1			1	5.3			1	5.3	16	95.6	47	177.1
1975	18	24.3	17	45.3	2	5.0	2	11.1	3	6.7			47	349.0	89	441.4
1976	26	31.6	11	33.3	2	10.8	4	12.1	2	6.0			84	623.3	129	717.1
1977	12	8.6	5	18.6	11	43.2	7	40.6	3	10.5			72	547.4	110	668.8
1978	14	26.3	4	10.5	2	5.6	7	29.8	2	8.6			79	587.8	108	668.6
1979	14	20.9	4	12.0	5	27.5	4	18.5	3	11.7	3	11.5	54	381.3	87	483.4
1980	10	7.9	8	34.8	2	1.8	8	29.2	3	6.7	5	16.4	73	435.2	109	531.9
1981	9	15.8	6	16.3	1	6.0	11	40.3	4	13.3	12	66.8	72	511.0	115	669.5
1982	9	11.6	2	6.4	2	6.0	15	75.2	7	20.5			62	467.3	97	587.0
1983	16	11.9	2	6.7	9	42.8	4	13.8	9	17.3	3	22.3	78	529.6	121	644.4
1984	5	11.3	5	15.2	1	2.5	5	27.8	9	33.7	6	34.3	64	457.4	95	582.2
1985	14	16.2	7	25.9	3	11.1	8	41.6	7	14.7	5	35.7	68	486.7	112	631.8
1986	14	15.6	5	17.8	6	31.9	8	44.7	10	25.7			71	490.6	114	626.1
1987	6	7.1	8	21.3	4	30.7	1	3.3	10	23.7			72	496.4	101	582.4
1988	11	7.1	10	27.3	9	65.6	4	19.0	7	17.4			68	499.7	109	636.0
1989	4	7.0	3	8.2	12	68.2	2	15.0	9	20.8			73	533.6	103	652.8
1990	1	2.5	4	13.2	12	46.5	2	10.3	10	28.3	1	6.9	45	327.9	75	435.4
1991	3	2.8	7	24.3	12	59.6	7	18.9	11	42.5			68	483.1	108	631.2
1992	22	18.8	6	11.2	7	42.5	5	11.9	6	13.6	1	7.7	84	643.5	131	749.1
1993	8	18.9	4	14.4	11	77.1	5	18.5	4	13.6	2	14.5	49	366.5	83	523.5
1994	12	11.0	7	17.4	13	82.6	5	18.6	12	42.8			74	560.9	123	733.3
1995	11	14.0	10	29.0	9	49.0	6	28.0	5	14.0			51	365.9	92	499.9
1996			2	2.8			1	3.4			2	10.3			5	16.5
Totals:	263	337.5	166	518.1	137	721.7	122	536.7	136	391.8	41	231.8	1424	10240	2289	12977

The year-by-year statistics for flight operations in table A1 permit estimation of the program effectiveness. Excluding the start-up year FY1974 (fig. 9), the total number of research flights for FY1975 through FY1995 was 1,446; 97 percent of these were astronomy research flights. (Non-astronomy research flights were devoted to a variety of topics, e.g., imaging the reentry of the Space Shuttle.) The average number of research flights per year was 72.3.

The number of aborted research flights plus the number of cancelled research flights represent the unreliability. Unfortunately, only totals of aborted and cancelled flights for all categories were recorded. From these records, the number of planned research flights that were aborted and cancelled is estimated to be about 200, so the resulting reliability for research flights flown/planned is 1446/(1446 + 200) = 0.88.

APPENDIX B. KAO SCIENTISTS

Table B1 lists all 126 Principal Investigators (PIs) and the number of flights awarded them by peer review for a total of about 510 KAO observing programs. Typically each listed PI had several people (co-investigators, graduate students, or postdoctoral associates) supporting his or her investigation. Names of both co-PIs are listed where appropriate. The number of science team members participating in KAO investigations probably exceeded 600. Many of these have gone on to significant roles in other astronomy programs or missions.

Indicated in the table are (a) individuals working at Ames for more than a year during the KAO era, (b) 33 KAO science-instrument team leaders, (c) participants on the SOFIA development team, (d) scientists who have advised on the planning and development of SOFIA, and (e) leaders of SOFIA science-instrument development teams. Clearly a large component of the science support for SOFIA is derived from experienced KAO astronomers.

Besides the astronomers recognized in table B1, roughly 200 others—including many graduate students and postdoctoral researchers—participated in the development of instrumentation for airborne observations.

Table B2 lists some of the scientists whose careers included experience with airborne instrumentation and observations, and some of their subsequent important contributions in ground- and/or space-based astronomy, including leadership roles in the astronomical community. No matter their subsequent activities, they will all vouch for the value of their experiences in developing and using airborne instruments. Apologies are due to the no-doubt significant number of other scientists whose names should appropriately appear in this table.

Table B1. KAO Principal Investigators and Awarded Flights, FY1974–FY1995

P I	flights	P I	flights	P I	flights	P I	flights
Aitken, David	1	Gehrz, Bob; d	1	Larson, Hal; b, d	61	Smith, Howard/	
Allamandola, Lou; a	5	Goebel, John; a	6	Laureijs, Rene	1	Strelnitski, Vladimir	1
Auguson, Gordon; a	1	Goldsmith, Paul	2	Lester, Dan; a, d	8	Soifer, Tom; d	23
Aumann, George; b	12	Greenhouse, Matt; d	2	Lynch, Dave	2	Sprague, Ann	2
Baluteau, Jean-Paul; b	16	Gulkis, Sam	4	Madden, Sue; a	2	Stacey, Gordon; b, d	5
Becklin, Eric; b, c	16	Haas, Mike; a, c	14	Maloney, Phil	1	Storey, John	2
Beckwith, Steve: b	11	Hanel, Rudy; b	2	McGregor, Peter	2	Strom, Steve	2
Betz, Al; b, d	41	Harper, Al; b, d, e	95	Melnick, Gary; d	11	Stutzki, Jurgen; d	4
Bezard, Bruno	1	Harvey, Paul; b, d	81	Morris, Mark; d	6	Tegler, Steve/	
Bjoraker, Gordon	9	Harwit, Martin; b, d	25	Moseley, Harvey; b, d	43	Weintraub, Dave	1
Boulanger, Francois	1	Helou, George; d	4	Mumma, Mike	3	Telesco, Charlie; d	8
Bregman, Jesse; a, c	13	Herter, Terry; b, e	14	Myers, Mike	7	Thronson, Harley; d	16
Burton, Michael; a	1	Hildebrand, Roger; b, d	47	Ney, Ed; K	7	Tielens, Xander; a, c	14
Butner, Harold	2	Hilgeman, Ted; b	4	Noll, Keith	1	Townes, Charles; b, d	63
Campbell, Murray	2	Hoffman, Bill; b, d	25	Novak, Giles	2	Townes, Charles/	
Campins, Umberto	9	Hollenbach, Dave; a	9	Omont, Allain	7	Genzel, Reinhard; b, d	36
Chin, Gordon; b	1	Houck, Jim; b	49	Pendleton, Yvonne; a	1	Wannier, Peter	7
Churchwell, Ed	3	Hughes, L.	2	Petuchowski, Sam	3	Waters, Joe; b	13
Cobb	1	Hunten, Don	2	Phillips, Tom; b	38	Weaver, Harold	5
Cohen, Martin	11	Hyland, Harry	4	Pipher, Judy; b, d	26	Wengler, Michael	2
Cox, Pierre/		Israel, Frank	3	Pollack, Jim; a	27	Werner, Mike; a, b, d	48
Tielens, Xander; a, c	1	Jaffe, Dan; d	16	Rank, Dave	2	Willner, Steve	7
Davidson, Jackie; a, c	6.5	Jones, Barbara; b	5	Rickard, Lee	12	Witteborn, Fred; a, b, d	47
DeGraauw, Thijs; b	6	Keene, Jocelyn	5	Roeser, Hans-Peter; b, c	17	Woodward, Chick	1
Dinerstein, Harriet; a	9	Klein, Mike	2	Rubin, Bob; a	5	Woodward, Chick/	
Dunham, Ted; a, b, c, e	3	Kleinman, Susan	8	Rudolph, Alex; a	3	Gehrz, Bob; d	1
Elliot, Jim; b, d	18	Knacke, Roger	8	Russell, Ray; b	6	Woolfe, Nick	1
Elvis, Martin	2	Knapp, Jill; d	1	Saykally, Richard	2	Wright, Ned	6
Erickson, Ed; a, b, c, e	94	Kuiper, Tom	8	Scoville, Nick	2	Yusef-Zadeh, Farhad	1
Evans, Neil/		Kutner, Mark	1	Simon, Mike	2	Zmuidzinas, Jonas; b, d, e	13
Mundy, Lee; d	23	Lane, Adair	2	Skinner, Chris	2	Zuckerman, Ben	2.5
Gautier, Nick	5	Langer, Bill	1	Smith, Howard	3		

a: resident at NASA Ames more than 1 year during the KAO era; b: KAO science-instrument team leader; c: SOFIA development team member; d: advisor on SOFIA planning; e: selected SOFIA science-instrument team leader.

Table B2. Some Participants in Airborne Instrument Developments and Some Subsequent Contributions		
Scientist	Current Affiliation	Notable Activities
Eric Becklin*	UCLA, retired/USRA	SOFIA contract Chief Science Advisor, former IRTF Director, HST/NICMOS instrument team
Steve Beckwith*	U. California.	Vice President for Research; former Director, STScI, MPIA
John Carlstrom	U. Chicago	Director, Kavli Institute for Cosmological Physics
Jackie Davidson	U. Western Australia	Former SOFIA contract Project Scientist
Thijs DeGraauw	ALMA	Director emeritus
Jessie Dotson	NASA ARC	NASA Ames Astrophysics Branch Chief
Darren Dowell	Caltech	SHARC photometer for CSO
Mark Dragovan	JPL/Caltech	CARA/PYTHON CMBR SI Team
Ted Dunham*	Lowell Observatory	PI SOFIA/HIPO; Kepler camera feasibility team
Jim Elliot*	MIT	SOFIA/HIPO Team
Ed Erickson*	NASA ARC, retired; Orbital Sciences Corp.	Original SOFIA Project Scientist for NASA; HST/NICMOS SI Team
Ian Gatley	RIT	Dean of Science
Reinhard Genzel*	MPE, Garching DE	Director; Herschel/PACS Team
Thijs de Graauw*	SRON, Groningen NL	Director, ALMA; PI Herschel/HI-FI, ISO/SWS
Matt Greenhouse	NASA GSFC	Project Scientist for JWST SI Payload
Mike Haas	NASA ARC	Director, Kepler Science Office
D. A. Harper*	U. Chicago	PI SOFIA/HAWC; former director CARA
Paul Harvey*	University of Texas	Mission Scientist, Herschel
Martin Harwit*	Cornell U., Emeritus	Mission Scientist, Herschel and ISO; SWAS Team
Terry Herter*	Cornell University	PI SOFIA/FORCAST; Spitzer support
Roger Hildebrand*	U. Chicago, retired	Former Astronomy and Astrophysics Department Chair
Jim Houck*	Cornell University	PI Spitzer/IRS; IRAS Co-I
Dan Lester	University of Texas	PI for SAFIR Vision Mission Study
Frank Low*	Infrared Laboratories	IRAS Co-I, Initial Spitzer Facility Scientist
Suzanne Madden	CEA Saclay FR	Herschel/SPIRE, PACS and SPICA/SAFARI SI teams
Gary Melnick	Harvard SAO	PI SWAS, Deputy PI Spitzer/IRAC
Alan Moorwood*	ESO	ESO Instrument Program Director
Harvey Moseley*	NASA GSFC	Detector systems for SOFIA/HAWC, Chandra, JWST
Giles Novak	Northwestern U.	Polarimeters SPARO for South Pole; SHARP for CSO
Tom Phillips*	Caltech	Director, CSO; U.S. team leader on Herschel
Judy Pipher*	U. Rochester, retired	Spitzer/IRAC Team
Albrecht Poglitsch	MPE, Garching DE	PI SOFIA/FIFI-LS and Herschel PACS
Tom Roellig	NASA ARC	Deputy SOFIA Project Scientist for NASA; Spitzer Facility Scientist, IRTS/MIRS (JP) SI team

Table B2. Concluded		
Hans-Peter Roeser*	U. Stuttgart DE	Managing Director, Institute for Space Systems
Michael Skrutskie	U. Virginia	PI, Two Micron All Sky Survey
John Storey	UNSW (AU)	Chair, School of Physics
Jűrgen Stutzki	U. Köln	Director, KOSMA; Co-Investigator SOFIA/GREAT
Tom Soifer	Caltech	Director, Spitzer Science Center
Charlie Telesco	U. Florida	Project Scientist, T-ReCS on Gemini South, CanariCam on Gran Telescopio Canarias
Xander Tielens	Leiden U. (NL)	Project Scientist Herschel/HI-FI
Alan Tokunaga	NASA IRTF Hawaii	Director
Charles Townes*	UC Berkeley, retired	PI, Ground-based Infrared Spatial Interferometer
Mike Werner*	JPL/Caltech	Project Scientist, Spitzer
Stan Whitcomb	LIGO/Caltech	Chief Scientist
Fred Witteborn*	NASA ARC, retired; Orbital Sciences Corp.	Original SIRTF (Spitzer) Project Scientist; Kepler camera feasibility team
Ned Wright	UCLA	Project Scientist, WISE
Jonas Zmuidzinas*	Caltech	PI SOFIA/CASIMIR; Herschel/HI-FI instrument team

* SI team leader on the KAO and/or Learjet Observatory

APPENDIX C. AN IMAGE HISTORY OF THE KAO

This appendix is a pictorial history of the Kuiper Airborne Observatory. Following images of the two icons of airborne infrared astronomy, Gerard Kuiper and Frank Low, the photos span the KAO's lifetime from the unmodified aircraft in 1971 until its decommissioning in 1995. Added at the end are three related images from 2008, 2010, and 2011. Ordered in chronological progression by year, these images trace some significant evolutionary aspects of the program. This graphic record is intended to recall the participation of many of its contributors, and to give the reader a sense of the character of the operation: its hands-on nature, its breadth of involvement by the science community, and its hospitality to young researchers and educators.

Nearly all the individuals pictured were involved with the KAO either as staff (ground crew, flight crew, and mission team) or as members of investigator teams. Many of the images were obtained from the Ames photo archive, others from individuals, and some from the authors. The latter tend to skew the distribution of investigator participants shown to include more from NASA Ames. This unintended bias is due to the availability of the pictures. The authors regret that images of many participants were not obtained and apologize for any unintended errors in references to those cited.

The year in which a photo was taken is shown when known; otherwise an estimated year is provided. Participants' affiliations given are those concurrent with the epoch of the photo. Students and postdoctoral associates are identified as far as possible. Instrument team leaders at the end of the program are listed in table 8. Principal Investigators are listed in table B1, and some participants are listed in table B2 with some of their subsequent professional activities. Some of those awarded significant recognitions by the science community are noted in table 13. Many of the investigators pictured are members of the American Astronomical Society (AAS). Their current affiliations can be obtained from the AAS public directory at http://members.aas.org/directory/public_directory_submit.cfm

Many of the images depict activities characteristic of the KAO operation, mirroring the descriptions in the body of this paper. The participants' expressions captured reflect a range of attitudes, from focused intensity, to acquiescence to being photographed, to lighthearted enjoyment, to results-inspired elation. Their legacy is the memories and accomplishments of the unique KAO program at NASA Ames, as well as the provision for the future of airborne astronomy: SOFIA.

Figure C1. Gerard P. Kuiper checking optical alignment of his instrument on the Convair 990 (Galileo), circa 1966.

Figure C2. Nancy Boggess and Frank Low with the Lear Jet, following observations of the luminous infrared object VY Canis Majoris, circa 1971. Pioneering airborne observations led by Kuiper (fig. C1) and by Low provided the rationale for building the KAO. Boggess managed NASA's infrared astronomy program, which included the KAO, from about 1977 until 1987.

1971

Figure C3. The C-141 Lockheed Starlifter, which became the KAO, probably at the Martin Marietta Georgia facility, 1971. It was one of the first C-141s built, a demonstration model. Lockheed produced about 285 of these aircraft for the U.S. Air Force.

1971

Figure C4. Bob Krouse (NASA Ames) with the KAO telescope, 1972. The telescope was assembled and tested at Ames before being shipped to Lockheed Aircraft Systems in Ontario, California, where the aircraft was modified and the telescope installed.

Figure C5. The KAO in 1974. Science observations began that year, with much engineering and performance evaluation accomplished on science flights. It was an exciting time in the program, and an exhilarating experience for scientists and staff alike.

1974

Figure C6. Graduate student Charlie Telesco (University of Chicago) and Bob Lowenstein (Yerkes Observatory), 1974.

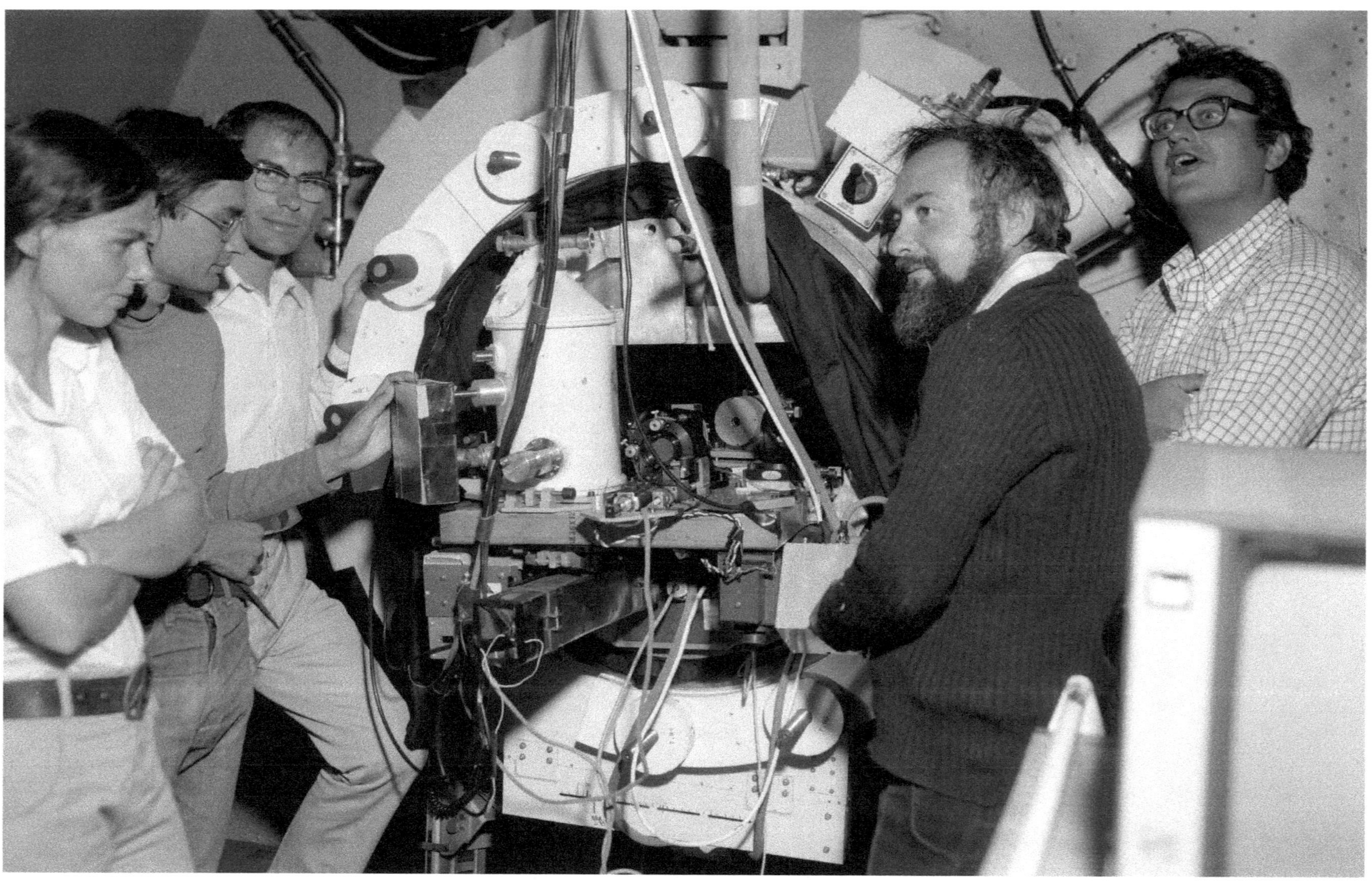

Figure C7. Therese Encrenaz and Jean Lecacheux (Paris Observatory), Guy Michel (Laboratoire Aimé Cotton, Orsay), Daniel Gautier and Yves Zeau (Paris Observatory), 1974. The French group used their Fourier Transform Spectrometer to observe Jupiter.

1974

Figure C8. Alan Moorwood (European Space Research and Technology Center (ESTEC), Netherlands), Jean Paul Baluteau (Observatoire de Meudon, France), John Beckman (ESTEC), Ezio Bussoletti (Observetoire de Meudon), Michel Anderegg (ESTEC), 1975, with Baluteau's high-resolution far-infrared Fourier Transform Spectrometer.

Figure C9. Rear: Dave Goorvitch, Al Ragasa; front: Fred Witteborn, John Gerdts, Ed Erickson, Jan Simpson, Larry Caroff, Gene Bekstrom, and Don Strecker (NASA Ames), 1975. Members of the Ames group, including others not pictured here, supported observations with a near-infrared circular variable filter spectrometer and a far-infrared Michaelson interferometer, both of which they had flown on the Lear Jet Observatory.

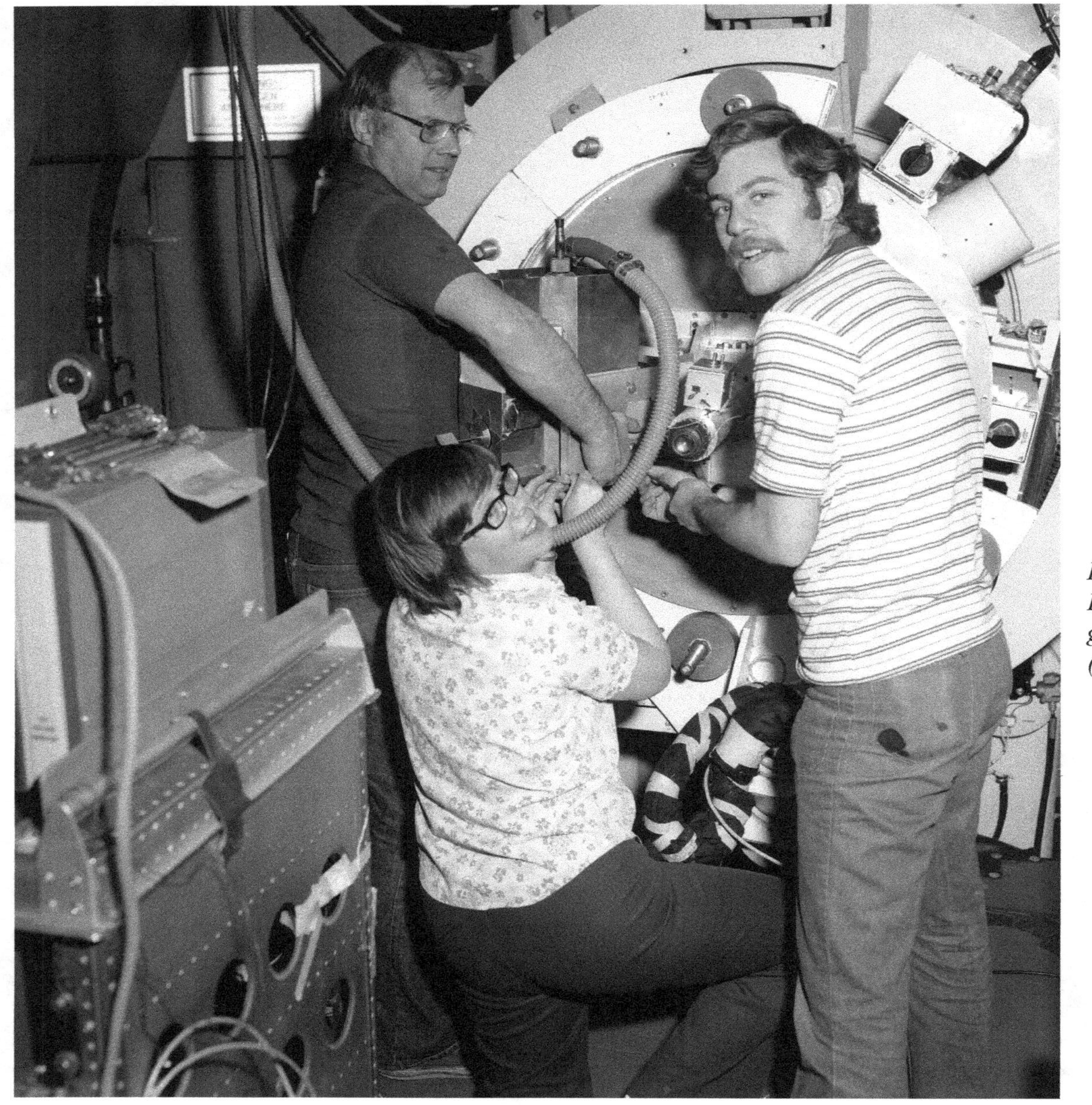

Figure C10. Professors Graeme Duthie and Judy Pipher, and graduate student Jerry Krassner (University of Rochester), 1976.

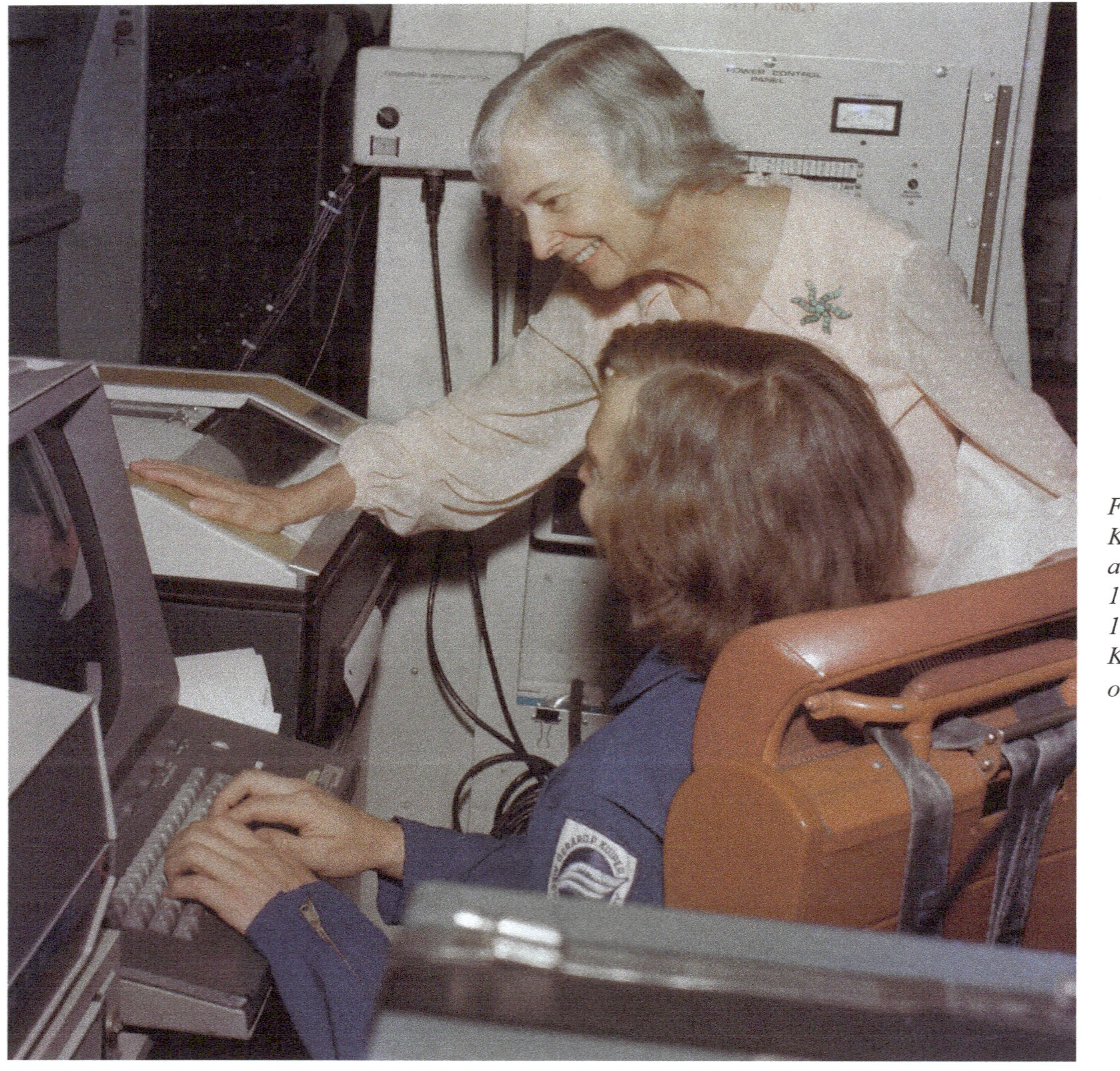

Figure C11. Mrs. Sarah Fuller Kuiper, Gerard Kuiper's wife, and Tom Mathieson (KAO crew), 1976. Gerard Kuiper died in 1973. Mrs. Kuiper attended the KAO dedication at NASA Ames on May 21, 1975.

Figure C12. Graduate student Ian Gatley, Eric Becklin, Gerry Neugebauer, and Gordon Forrester (Caltech), 1976, making in-flight repairs to their far-infrared photometer.

Figure C13. Lloyd Domier and Louie Russo (NASA Ames), 1977. Domier was the aircraft crew chief and Russo was a member of the ground crew for the KAO throughout its operational lifetime at Ames. Ground crew members were all NASA civil servants.

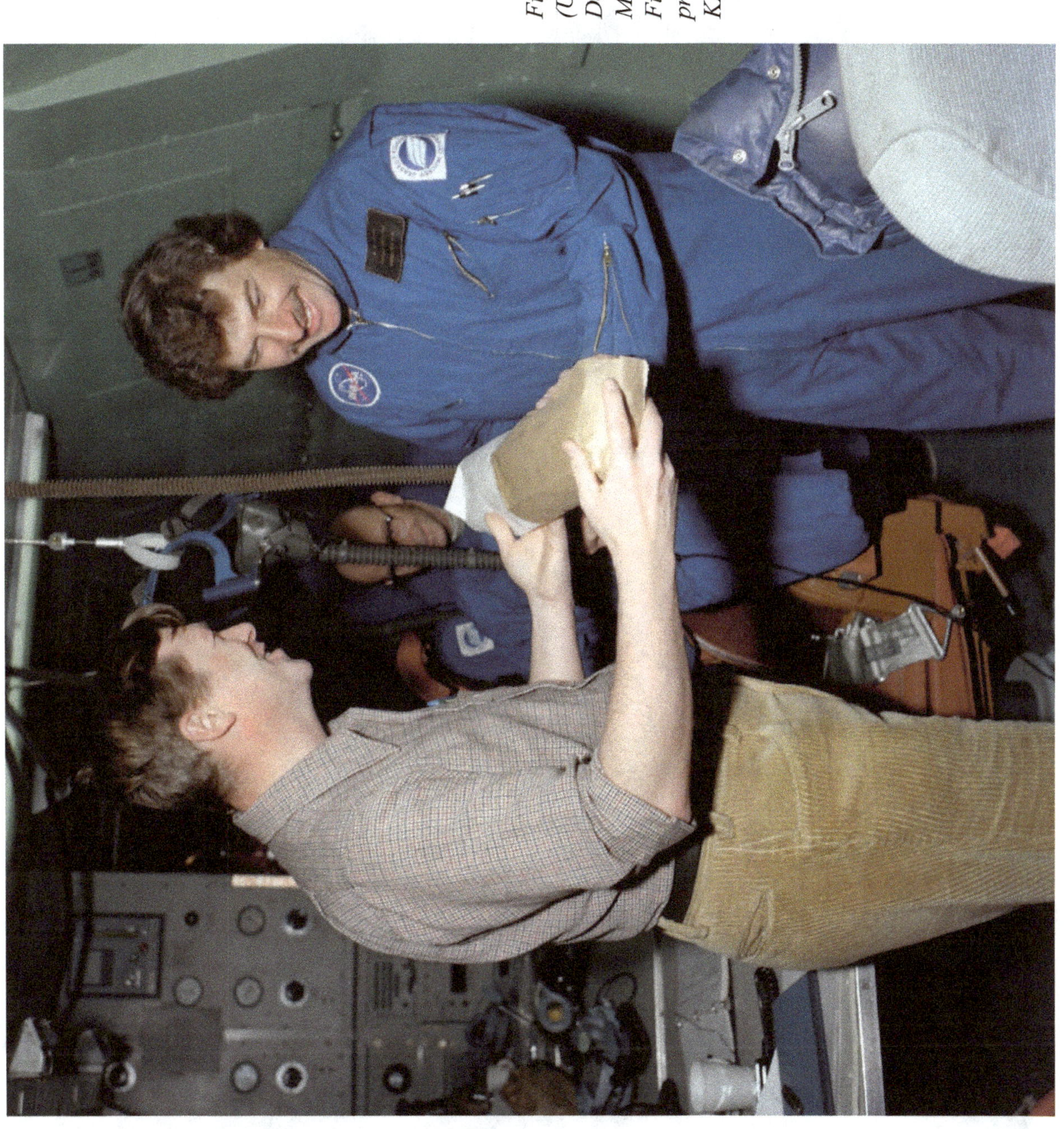

Figure C14. Uwe Fink (University of Arizona), and Don Olsen and Jim McClenahan (KAO staff), 1977. Fink is presenting a preflight present for McClenahan and the KAO crew.

Figure C15. Richard "Tex" Ritter (L) and Dave Barth (R), both NASA civil servants, 1977, in the three-man cockpit of the C-141. Ritter is flying as copilot and Barth as the flight engineer on this mission. Barth was killed and Ritter was permanently injured when a NASA helicopter they were operating crashed in 1981.

Figure C16. Milo Reisner, Don Olson, and Jim McClenehan (NASA Ames) at the telescope console 1977.

Figure C17. Tom Wilson (KAO Staff) and Pete Kuhn (Raven Systems and Research), 1977. Kuhn developed an infrared radiometer intended to detect clear air turbulence. He tested the instrument on many KAO flights.

Figure C18. Kevin Krisciunis and Tom Wilson (Ames computer staff), and Richard Poppen (University of Arizona), 1977.

Figure C19. Bill Rickets, Paul Swanson, and Tom Kuiper (Jet Propulsion Laboratory) 1977. The instrument is a 187-GHz receiver built for Joe Waters' atmospheric research, but adapted here for astronomical observations.

Figure C20. Paul Swanson (Jet Propulsion Laboratory (JPL)), Pat Thaddeus (Goddard Institute for Space Studies; Columbia University), and Joe Waters (JPL), 1977.

Figure C21. Patrick Atchison (NASA Ames) and Daryl Mason (Travis Air Force Base), 1977. Mason was a C-141 aircraft mechanic. The Air Force, which operated a fleet of C-141s, provided much of the maintenance on the KAO aircraft.

Figure C22. Jim Elliot, graduate student Ted Dunham, and Doug Mink (Cornell University), making preparations to observe the occultation of a star by Uranus, 1977. Their work yielded unexpected success—they discovered the rings of Uranus from the KAO, flying from Perth Australia over the Indian Ocean.

Figure C23. Graduate student Kris Sellgren and Mike Werner (CalTech), 1977, checking out instrumentation to measure luminosities and morphologies of HII regions (gas clouds ionized by hot stars) using the Caltech far-infrared photometer.

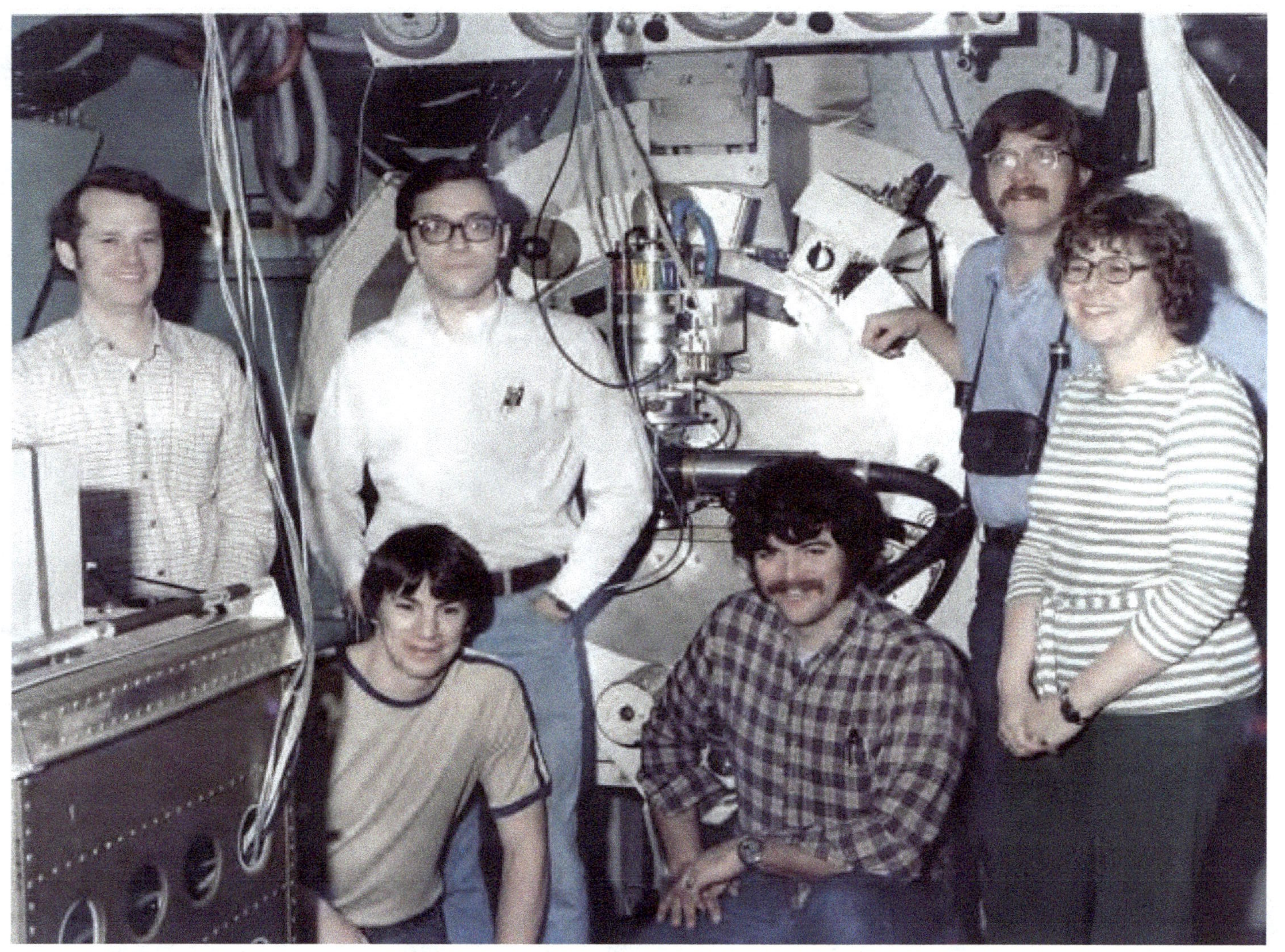

Figure C24. Jim Houck and Dan Briotta (Cornell University), graduate student Terry Herter (kneeling) (University of Rochester), George Gull (CU), Bill Forrest, Judy Pipher (U of R), with the Cornell mid-infrared grating spectrometer, 1977. Houck's group also observed earlier from the Learjet, as did Forrest.

Figure C25. Greg Buck (L) (NASA Ames) and Roger Knacke (R) (State University of New York), 1977. Buck, a technician, and Knacke, a professor on sabbatical, are preparing for flights of the dual far-infrared Michaelson interferometers from Ed Erickson's group.

1977

Figure C26. Ben Horita and Don Oishi (NASA Ames), 1978. They are in the telescope cavity, preparing to remove the telescope from the C-141 for maintenance.

Figure C27. Don Oishi, (NASA Ames), 1978. The telescope is supported with adjustable rigging in preparation for removal from the plane.

Figure C28. Far left: Barry Turney (face behind chain), and Louise Gorgonzola-Nitz behind Lloyd Domier. On floor: Bob Walker, Don Oishi, Bruce Kelly, and Milo Reisner behind Ben Horita (standing at right); behind telescope: Fred Pants, (NASA Ames), 1978. This was the first time the telescope was removed since its original installation in the plane in 1972. In addition to participating in telescope maintenance, Oishi, Kelly, Reisner, and Horita all flew as telescope operators.

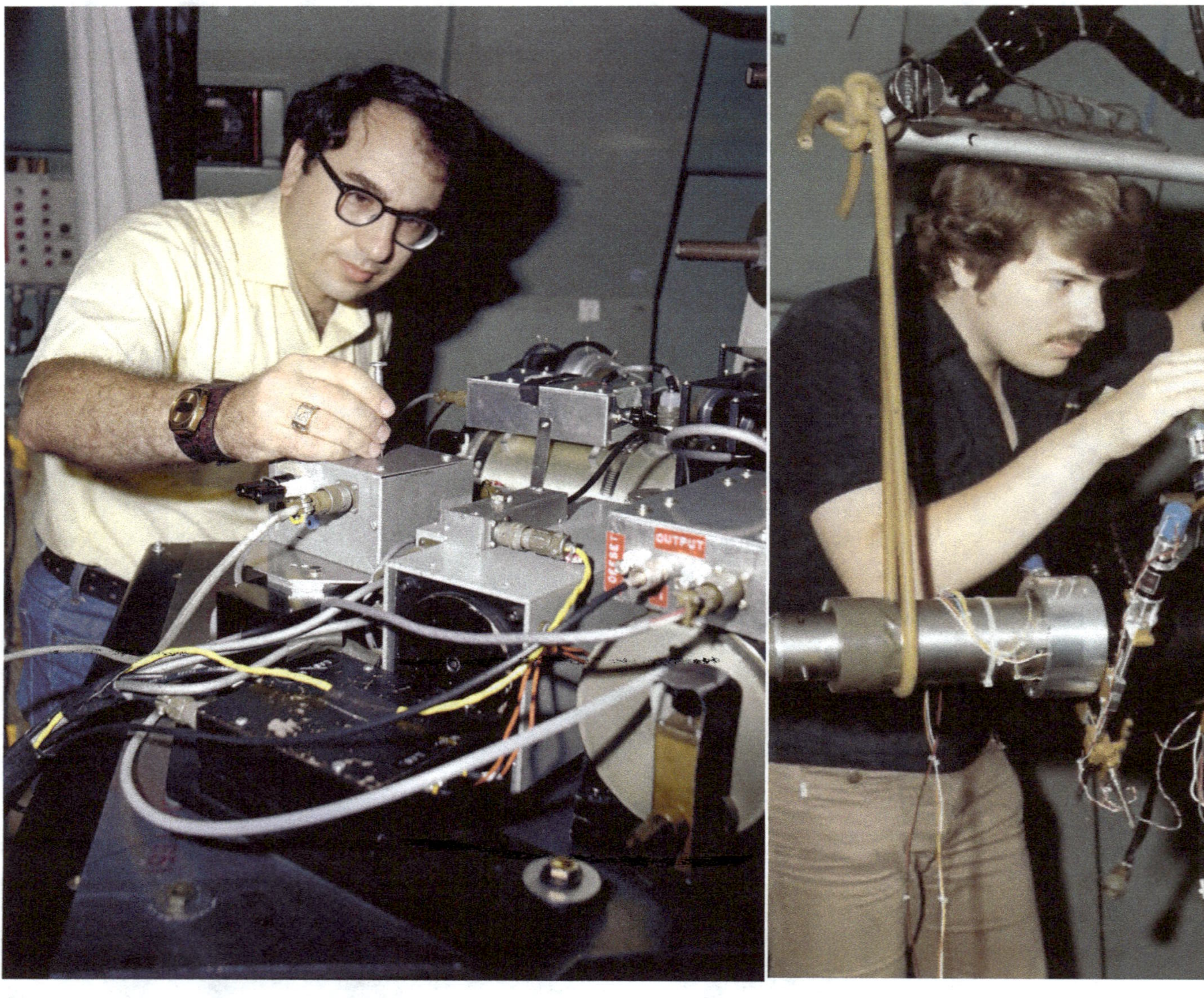

Figure C29. Ted Hilgeman (L) (Northrop Grumman Corporation) and Rick Puetter (R) (University of California at San Diego), 1978. In the early days of the KAO operation, relatively small groups of scientists and engineers developed state-of-the-art science instruments.

1978

Figure C30. Charles Townes, with graduate student Sara Beck (U.C. Berkeley), Tom Matheson (KAO computer operator), and graduate student Don Brandshaft (U.C. Berkeley) monitor incoming data from the UCB Fabry Perot far-infrared spectrometer, 1978. Townes' students flew an earlier version of the spectrometer on the Learjet.

Figure C31. Members of the Infrared Imaging-of Shuttle (IRIS) team installing their equipment, 1979. IRIS obtained thermal images of the space shuttle as it reentered the Earth's atmosphere, in an effort to measure the distribution of temperatures on the shuttle surfaces. It was a challenging problem to position the KAO and the telescope to catch the brief events.

Figure C32. Graduate students Lee J. Rickard and Harley Thronson (University of Chicago), 1979. Thronson was one of the first students to earn his doctorate using data obtained on the KAO. He was a member of Al Harper's team.

Figure C33. Al Harper's group (Yerkes Observatory/University of Chicago) on the KAO, 1979. Ed Shaya, Al Harper, Jocelyn Keene, Harvey Moseley, Bob Lowenstein, and Jim Smith. Harper earned his Ph.D. with Lear Jet observations made with Frank Low. He observed from the KAO throughout its lifetime, mentoring numerous graduate students including Shaya, Keene, and Moseley.

Figure C34. Steve Willner (University of California at San Diego), Tom Soifer (Caltech), and KAO telescope technician Bob Walker, 1979. The UCSD 4- to 8-micron circular variable filter spectrometer they used detected unidentified 6.2- and 7.7-micron features, now known as polycyclic aromatic hydrocarbons (PAH) molecules.

Figure C35. Judy Pipher, Larry Helfer, and graduate student Terry Herter (University of Rochester), 1979, with the Cornell (not Hawaii) mid-infrared grating spectrometer.

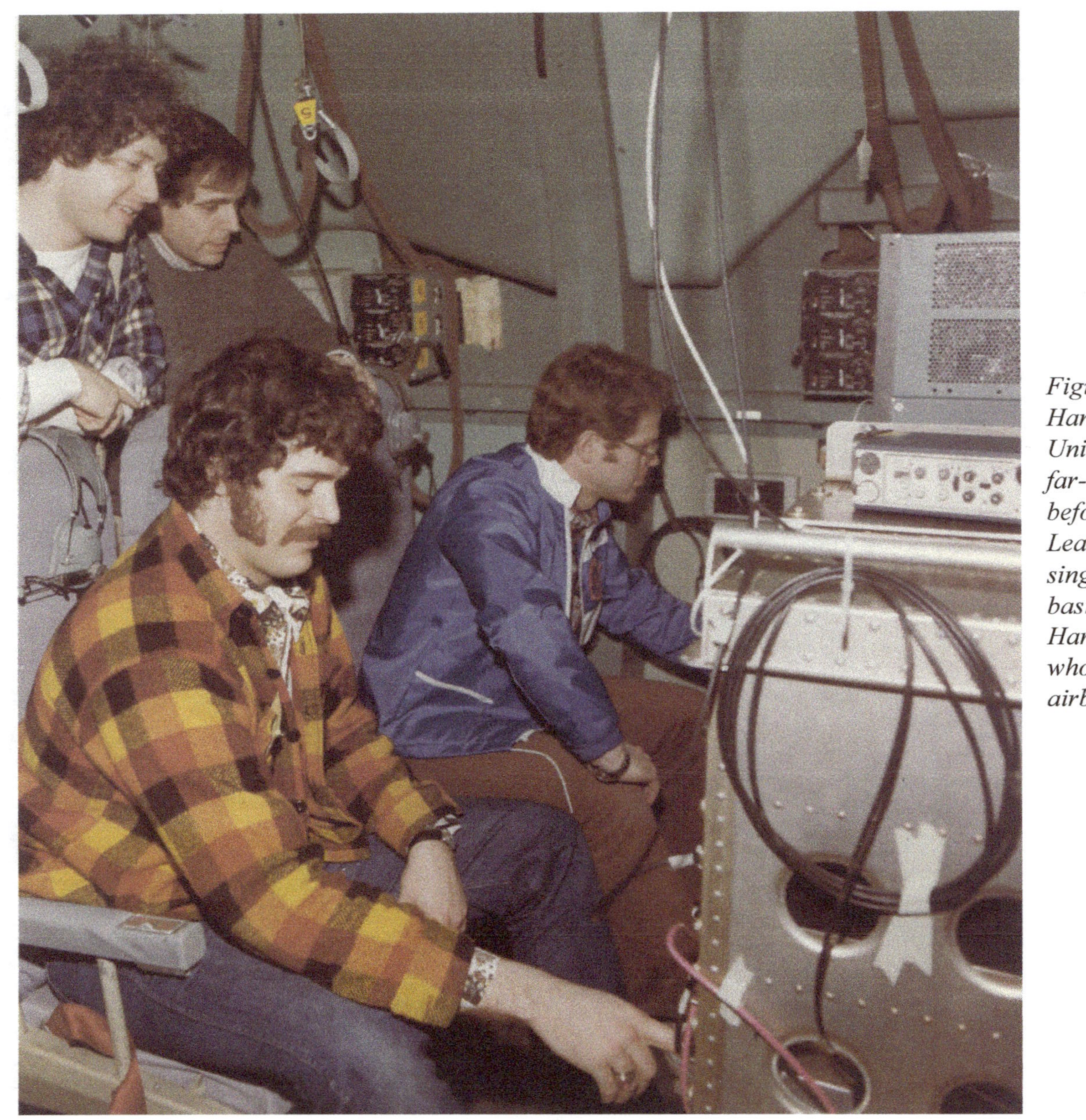

Figure C36. Gary Melnick (student), Martin Harwit, George Gull, and Ray Russell (Cornell University), 1979. Harwit and his team did far-infrared spectroscopy from the Learjet before the KAO became operational. From the Lear, they first detected the 158-micron line of singly ionized carbon, which characterizes basic processes in the Interstellar Medium. Harwit also mentored a number of students who earned their doctorates for work on the airborne observatories.

Figure 37. Larry Caroff, Ed Erickson, and Darrell McKibben (NASA Ames) monitor incoming data displayed on a strip chart, 1979.

Figure C38. KAO staffer Bruce Kelly and Thijs DeGraauw, (Netherlands Institute for Space Research, Gronigen), 1980, inspecting the installation of de Graauw's heterodyne receiver on the KAO telescope.

Figure C39. Howard Smith, Scott Davis, Bob Johnson, Guy Michel, Hal Larson, Mike Williams (University of Arizona), with Larson's high-resolution near infrared Fourier Transform Spectrometer, 1980. Preparations to install science instruments was sometimes done in a hallway of the KAO hangar because of limited lab space.

1980

Figure C40. Bob Lowenstein (Yerkes Observatory), Harvey Moseley (NASA Goddard Space Flight Center), Al Harper, graduate student Ed Shaya , and Roger Hildebrand (University of Chicago), 1980.

Figure C41. The KAO management team celebrating the first 5 years of full KAO operations: Jim McClenahan, Bob Cameron, Carl Gillespie, and Lou Haughney (NASA Ames), 1980. Cameron headed the Medium Altitudes Missions Branch that managed the KAO and other aircraft science programs. All flew as mission directors on astronomy flights.

Figure C42. KAO Software Development Team, 1980. Clockwise from left: Vince Bellows, Vic Loesche, Jeff Terry, Peter Fiekowski, Jim Panteleo, Kevin Krisciunis, Steve Culp, Tom Matheson, Sarah Young, and Randy Janke. Bellows and Janke worked primarily on the Convair 990 data systems. Krisciunis, Culp, Matheson and Young flew as computer operators on KAO astronomy missions.

Figure C43. Charles Townes and Reinhard Genzel (University of California at Berkeley), 1981. Here they are planning a mapping strategy for locating guide stars to be used on observations with Townes' far-infrared Fabry Perot spectrometer.

Figure C44. Judy Pipher (U. Rochester), George Gull, Dan Briotta, Mark Shure, Terry Herter, and Jim Houck (Cornell University), 1981. A number of students' Ph.D. theses included data from airborne observations—including Herter's (Pipher's student) and Shure's (Houck's student).

Figure C45. Graduate student Dan Watson and postdoctoral associate John Storey, from Charles Townes' group at University of California at Berkeley, 1981. Both Watson and Storey were deeply involved in the development of technology for Townes's Fabry Perot spectrometer.

1981

Figure C46. Jocelyn Keene and Chas Beichman with the Caltech heterodyne receiver of Tom Phillips, 1981. Science instruments were almost always updated or reconfigured between flight series, so that much effort was expended in adjusting them for optimum performance.

Figure C47. Tom McMahon (KAO staff Tracker Operator), and Mike Werner (JPL), circa 1981. Werner was using the hand-held "tweaker box" to update the telescope pointing while McMahon was having his in-flight lunch.

Figure C48. Tom Phillips, Jocelyn Keene, and Chas Beichman (California Institute of Technology), 1981, starting a transfer of liquid helium in Phillips' Heterodyne Receiver in the C-141 hangar.

Figure C49. Graduate students Mark Shure and Paul Graf (Cornell University), 1982. Both were mentored by Jim Houck and wrote their Ph.D. theses using data they obtained on observations from the KAO.

1982

Figure C50. John Brown and Bob Tinkey (NASA Ames), 1983. Brown and Tinkey were flight safety inspectors. All mission systems and science instrument installations were inspected for airworthiness before every flight.

Figure C51. Terry Stoeffler (L) and Chris Rhoades (R). Terry was the deputy crew chief for the C-141. Chris was a helicopter mechanic who assisted with preparations for expeditions of the KAO.

Figure C52. Ames aircraft mechanics Monte Hodges and Randy Hobbs, 1983. KAO ground crewmen were all NASA civil servants.

Figure C53. Barbara Jones and Steve Willner (University of California at San Diego), and Judy Pipher (University of Rochester), circa 1983. Collaborations between researchers from different institutions were based on their interests and the capabilities of the relevant science instrument.

1983

Figure C54. Roger Hildebrand and graduate students Mark Dragovan and Giles Novak (University of Chicago),1983. Hildebrand's group pioneered far-infrared polarimetry. Eleven of his students wrote Ph.D. dissertations using data they obtained on the KAO.

Figure C55. Jeremy Wyant, Sara Beck, graduate student Michael Skrutskie, undergraduate John Carr (Cornell Univeristy), Neal Evans (University of Texas at Austin), Antonella Natta (Arcetri Observatory, Italy), and Steven Beckwith (Cornell University), 1983. Beckwith's group developed and flew a medium-resolution mid-infrared grating spectrometer to study processes related to star formation in the interstellar medium. Evans and Natta were collaborating guest investigators, Beck a post-doctoral associate, and Wyant a group technician.

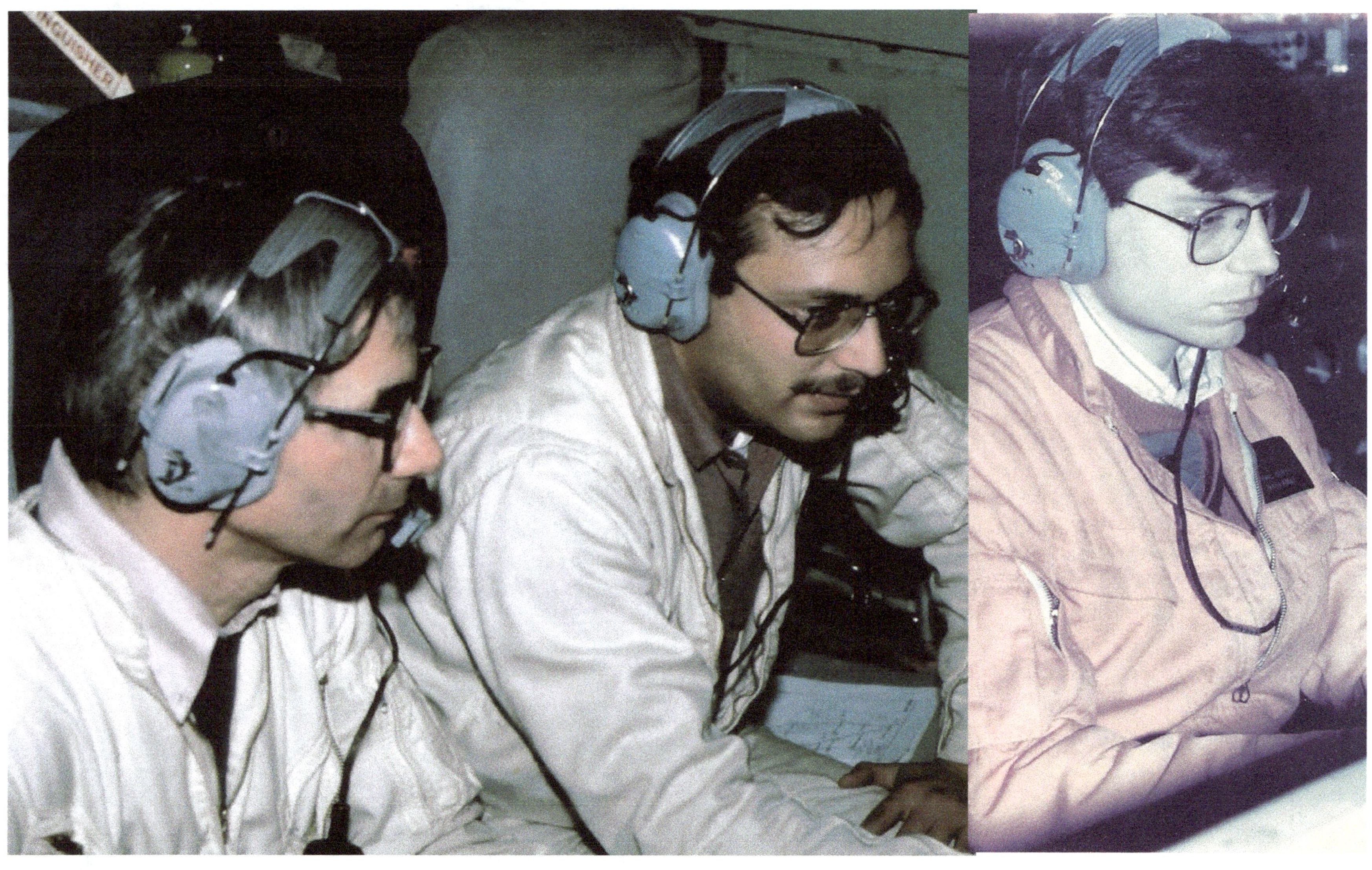

Figure C56. Ed Erickson, Juergen Wolf (NASA Ames), and Phil Duffy (Stanford University), circa1984. Wolf was a post-doctoral research associate from Germany, and Duffy a Stanford graduate student, both working in Erickson's group. Here they are monitoring the data stream from the Ames Cryogenic Grating Spectrometer. Duffy's dissertation was based on far-infrared spectral lines measured from the KAO.

1984

Figure C57. Undergratuate student Chuck Fuller, Gordon Stacey, Paul Viscuso, and Martin Harwit (Cornell University), 1984. A number of KAO investigators provided undergraduate students from their institutions exposure to the airborne astronomy operation.

Figure C58. Foreground: Jim Cooley and Bill Rose (Rose Engineering) with student Todd ?. Background: Jim Elliot and Ted Dunham (MIT), 1984. To understand effects of flight and cavity conditions on optical performance ("seeing"), Rose used the electronics in the rack to monitor pressure transducers in the cavity and on the telescope while Dunham and Elliot measured image quality.

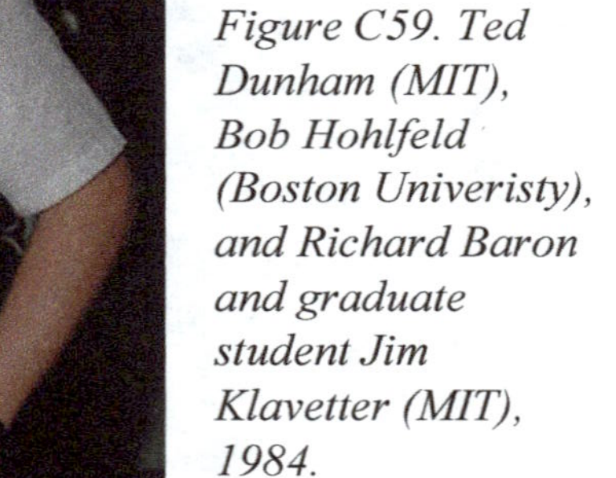

Figure C59. Ted Dunham (MIT), Bob Hohlfeld (Boston Univeristy), and Richard Baron and graduate student Jim Klavetter (MIT), 1984.

Figure C60. Rear: Mark Schaeffer, Herb Pickett, and Tom Kuiper; front: Paul Batelaan, Mike Klein (JPL), and Peter Zimmermann (U. Cologne); 1985. KAO instrument teams often included experts in cutting-edge technologies, such as Zimmermann, a developer of heterodyne receivers.

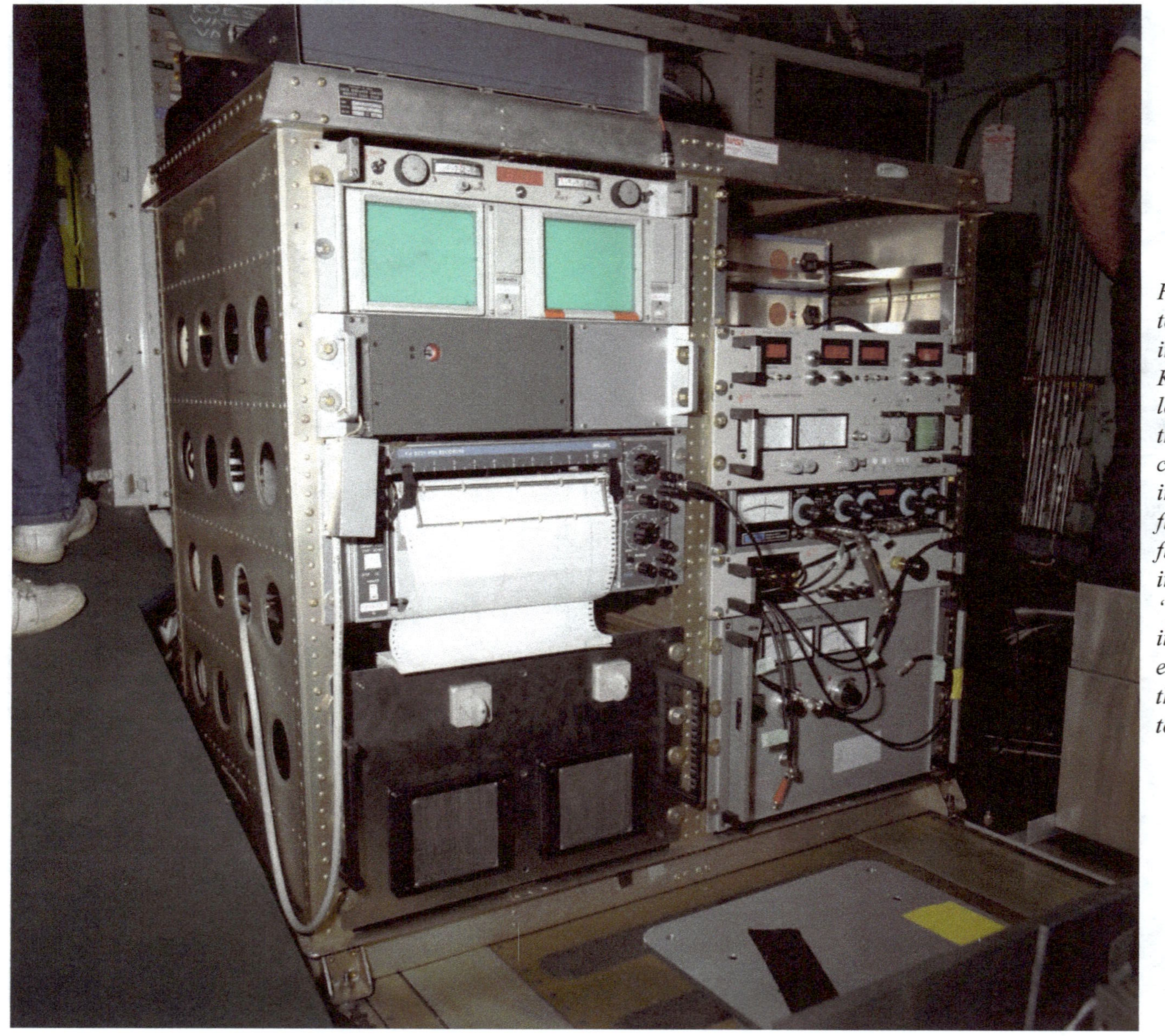

Figure C61. An instrument team's electronics rack installed in the back of the KAO, 1985. Each team was loaned an empty rack to hold their electronics, which they could install at their home institution before arriving for flights. A number of teams flew their equipment installed in the rear of the KAO in a "piggyback" mode, to verify its operation in the aircraft environment before mounting their instrument on the telescope.

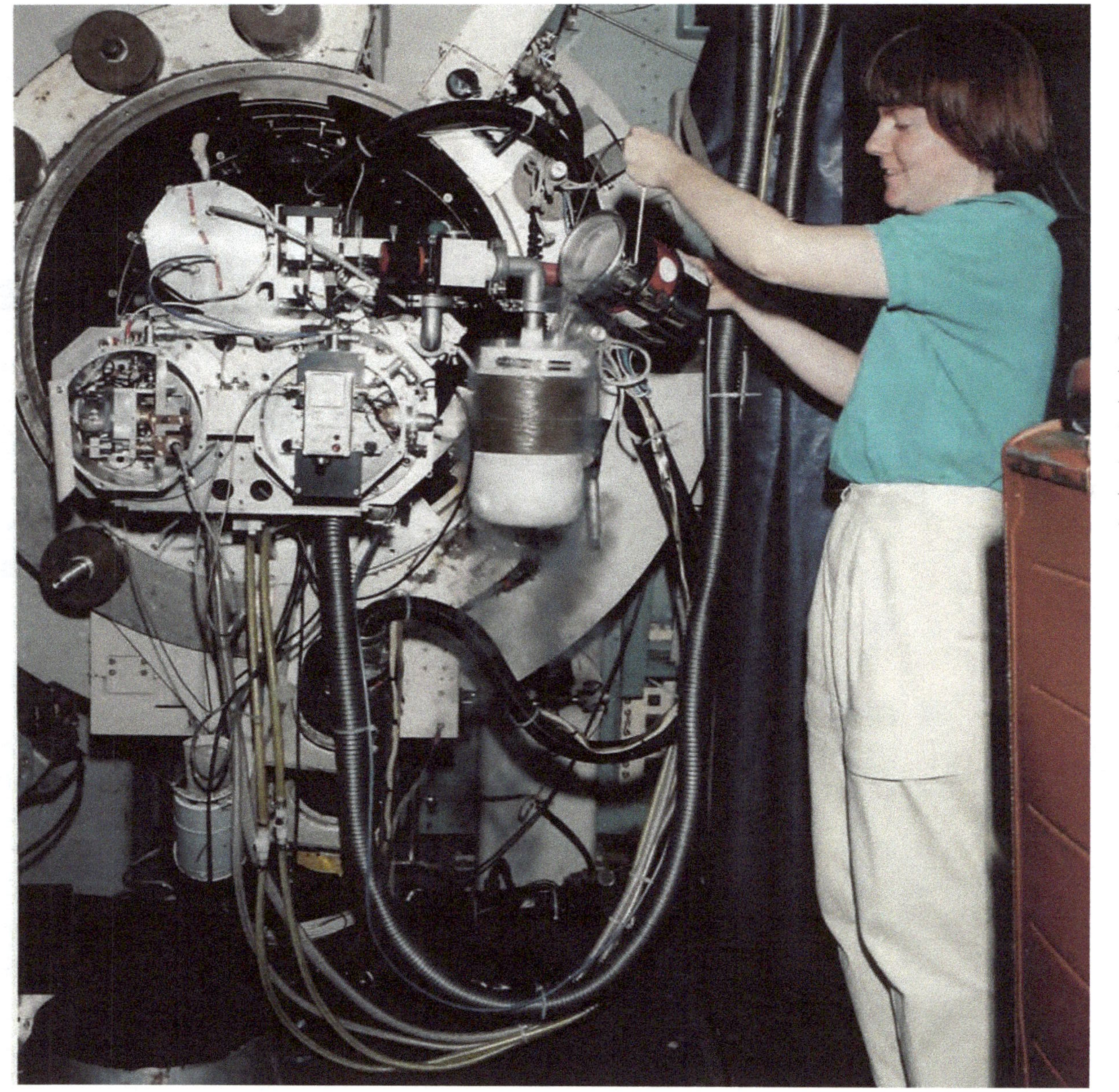

Figure C62. Post-doctoral associate Ruth Titz (Max Planck Institut for Radioastronomie, Bonn, Germany), 1985, resupplying the heterodyne receiver of Hans Peter Roeser with liquid nitrogen.

1985

Figure C63. Peter van der Wal, Hans Peter Roeser, Walter Esch, Roland Wattenbach, and Frank Schaefer (Max Planck Institut for Radio Astronomie, Bonn, Germany), 1985.

1985

Figure C64. High school student Lindsey Davenport, Mike Haas, Martin Burgdorf, Ed Erickson, and Sean Colgan, 1985, starting pre-flight checkout of the Ames Cryogenic Grating Spectrometer (CGS). Haas, Erickson, and Colgan were at NASA Ames; Burgdorf was a German post-doctoral associate in Erickson's group. A few KAO investigators mentored high school students, but they never flew on the KAO.

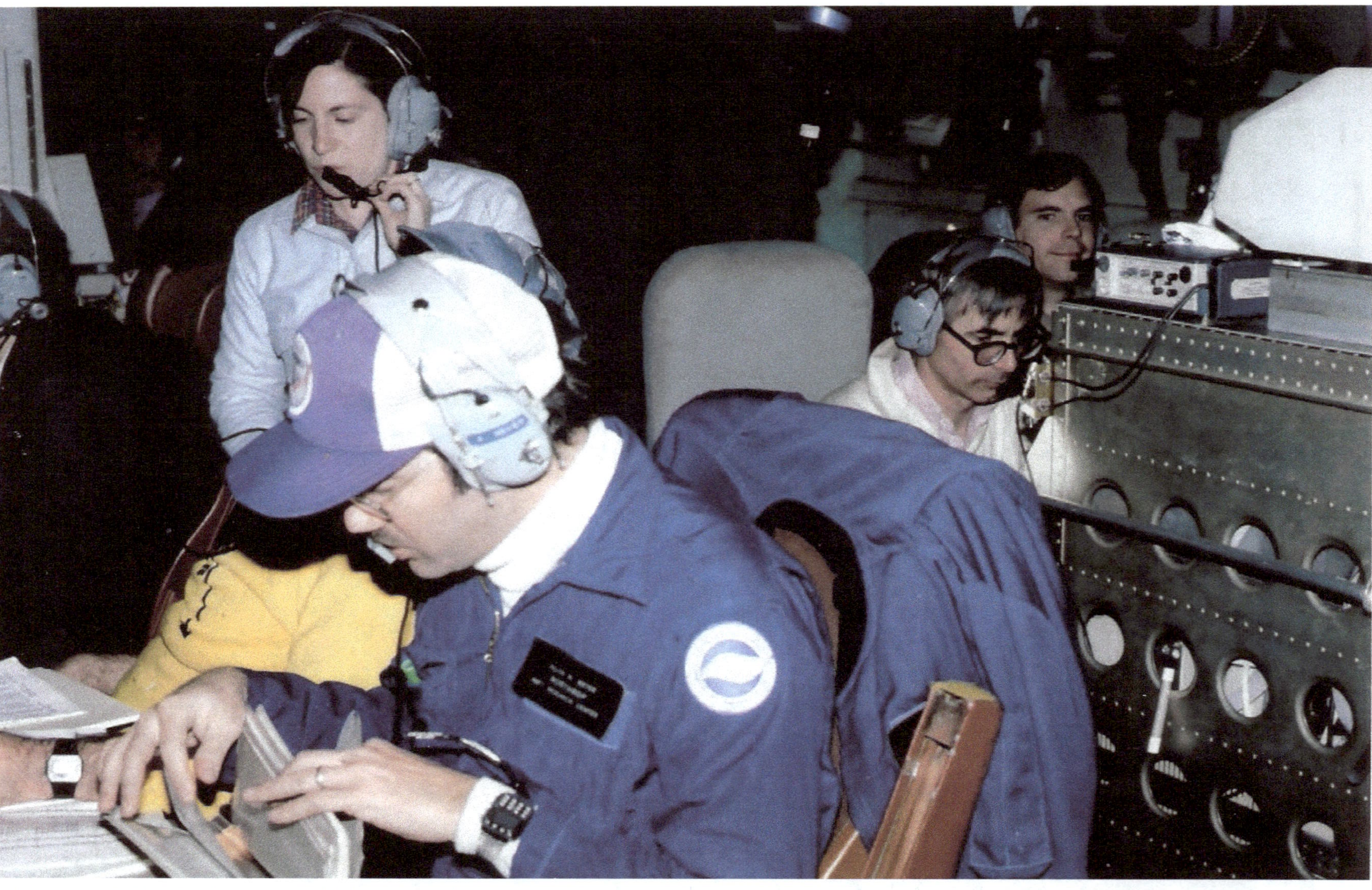

Figure C65. Harriet Dinerstein (University of Texas), Allan Meyer, Ed Erickson (NASA Ames), and Benton Ellis (University of Texas), circa 1985. Ellis was a postdoctoral associate working with Dinerstein; they were Guest Investigators working with Erickson's team.

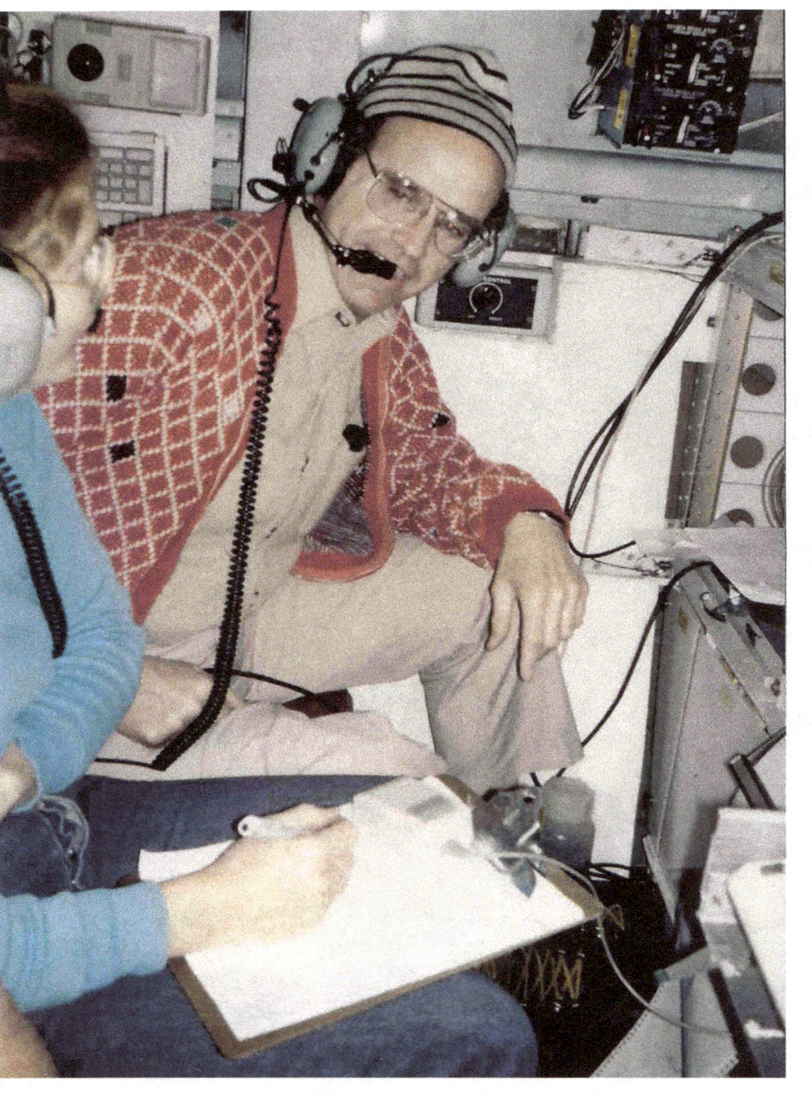

Figure C66. Diane Wooden and Lou Allamandola (L) and Xander Tielens (R) (NASA Ames), circa 1985. Allamandola's Astrochemistry Laboratory elucidated properties of ices and polycyclic aromatic hydrocarbons (PAHs). Tielens explored theoretically the physics and chemistry of the interstellar medium. Both pursued their interests as Guest Investigators on the KAO.

Figure C67. Some of the KAO NASA staff: Lou Haughney, Jim McClenahan, Don Oishi, Bob Walker, Milo Reisner, Ben Horita, and B. B. Gray, 1985. All devoted years of their careers to ensuring the success of the KAO.

1985

Figure C68. Rear: Carl Rice and Ray Russell (Aerospace Corporation), Mike Reichenbach and Dave Pollock (Teledyne Brown Engineering), and Dave Lynch (Aerospace Corporation); front: John Hackwell, Bob Macklin, and Dave Retig (Aerospace Corporation), and Robert Keever (Boeing). This team, led by Russell, was performing a technology evaluation. The few such non-astronomy research programs flown on the KAO typically reimbursed the program for their flight opportunities.

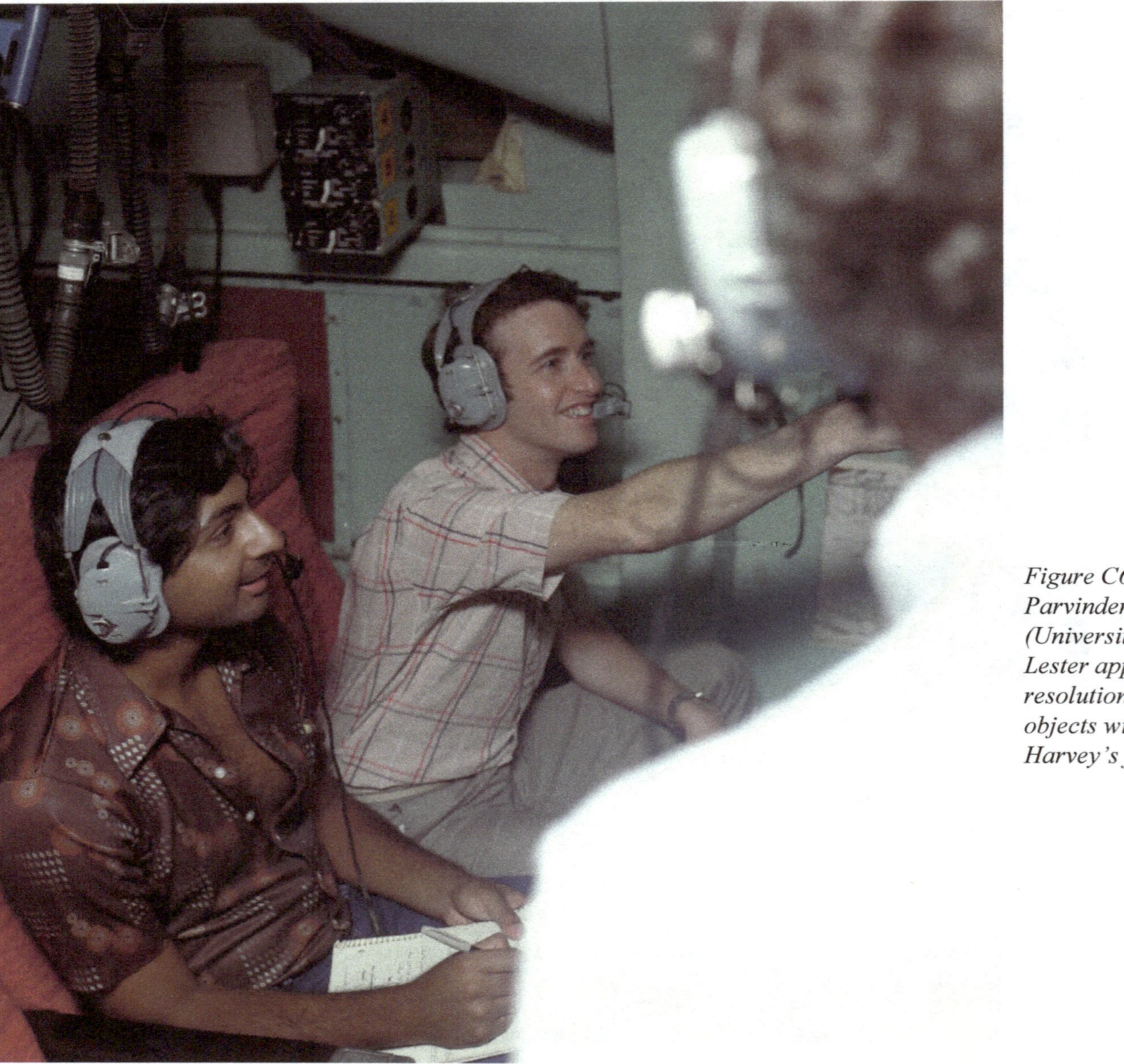

Figure C69. Graduate student Parvinder Parmer and Dan Lester (University of Texas at Austin), 1986. Lester applied the technique of super resolution to the study of young stellar objects with Paul Harvey, using Harvey's far-infrared photometer.

Figure C70. Paul Harvey (University of Texas at Austin), 1986. Harvey's group studied a variety of objects in the Interstellar Medium using his far-infrared photometer. Here he adjusts the telescope pointing at the Tracker Console. Harvey published more research papers based on KAO data than any other investigator.

Figure C71. Joanna Lees, Al Harper, David Hughes, Steve Most, and Darren Dowell (University of Chicago), circa 1986. Lees and Dowell were graduate students participating in flights of Harper's far-infrared photometer.

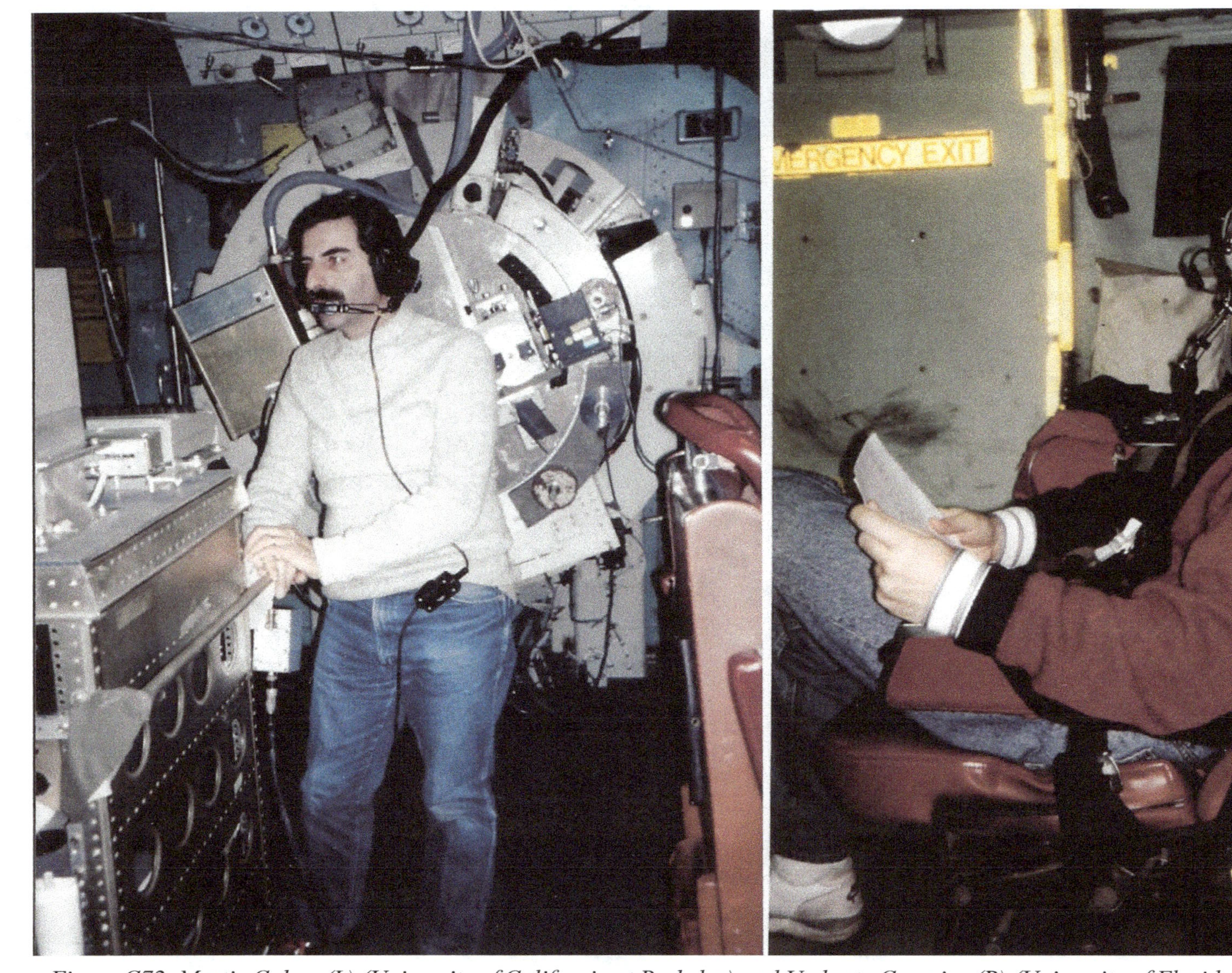

Figure C72. Martin Cohen (L) (University of California at Berkeley) and Umberto Campins (R) (University of Florida), circa 1986. They were "guest investigators," pursuing their separate interests by collaborating with science instrument teams.

Figure C73. Graduate student Jonas Zmuidzinas, Rita Boreiko, and Al Betz (University of California at Berkeley), 1987. Betz pushed the technology of heterodyne receivers to higher frequencies, eventually enabling the group to measure the velocity structure of the important 63-micron neutral oxygen line in Orion from the KAO.

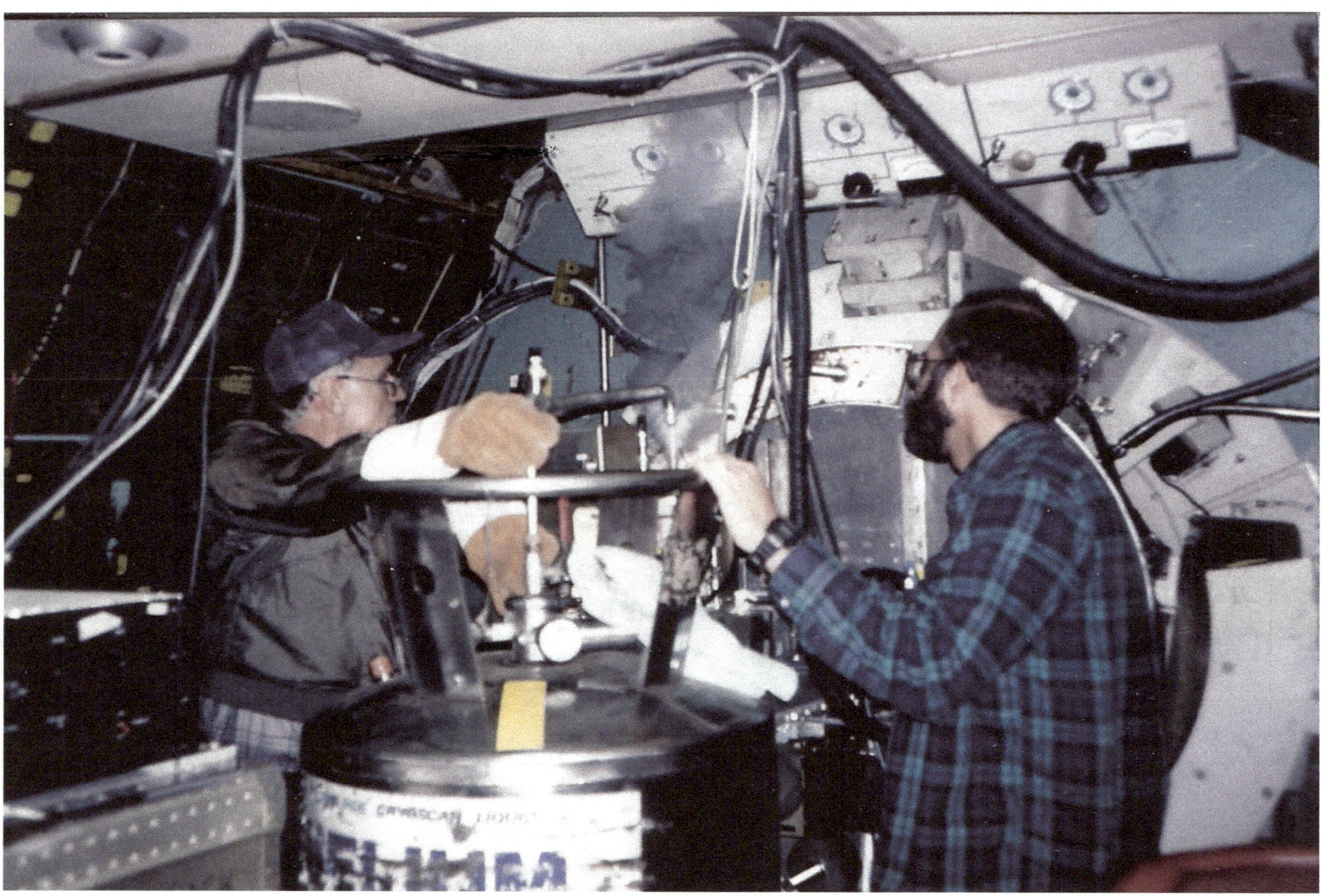

Figure C74. Fred Witteborn and Scott Sandford (NASA Ames), transferring liquid helium into Witteborn's Hi-Fogs mid-infrared spectrometer, circa 1987.

Figure C75. KAO and crew in Christchurch, New Zealand, 1987. From 1986 until 1995, the KAO deployed at least once every year to Christchurch, New Zealand, for observations in the southern hemisphere. This is a rare image because it includes members of the ground crew, flight crew, mission staff, and science teams.

Figure C76. Jan Simpson and Mike Haas (NASA Ames), circa1987, are elated to show a plot of a spectral line just measured with the Ames far-infrared Cryogenic Grating Spectrometer, aka the CGS.

Figure C77. Top: Team leader Tom Kuiper; middle: Herb Pickett, Peter Wannier (JPL), Patrick Dierich (Observatoire de Paris), Paul Batelann, and Mark Schaeffer (JPL); front: Laurent Denis (Observatoire de Paris) and Peg Frerking (JPL), 1987.

1987

Figure C78. Hal Larson, Keith Noll, Mike Williams, and Scott Davis (University of Arizona), 1988. Larson and his team discovered water in Jupiter's atmosphere and comets from the KAO, as well as obtaining significant results on a variety of other objects.

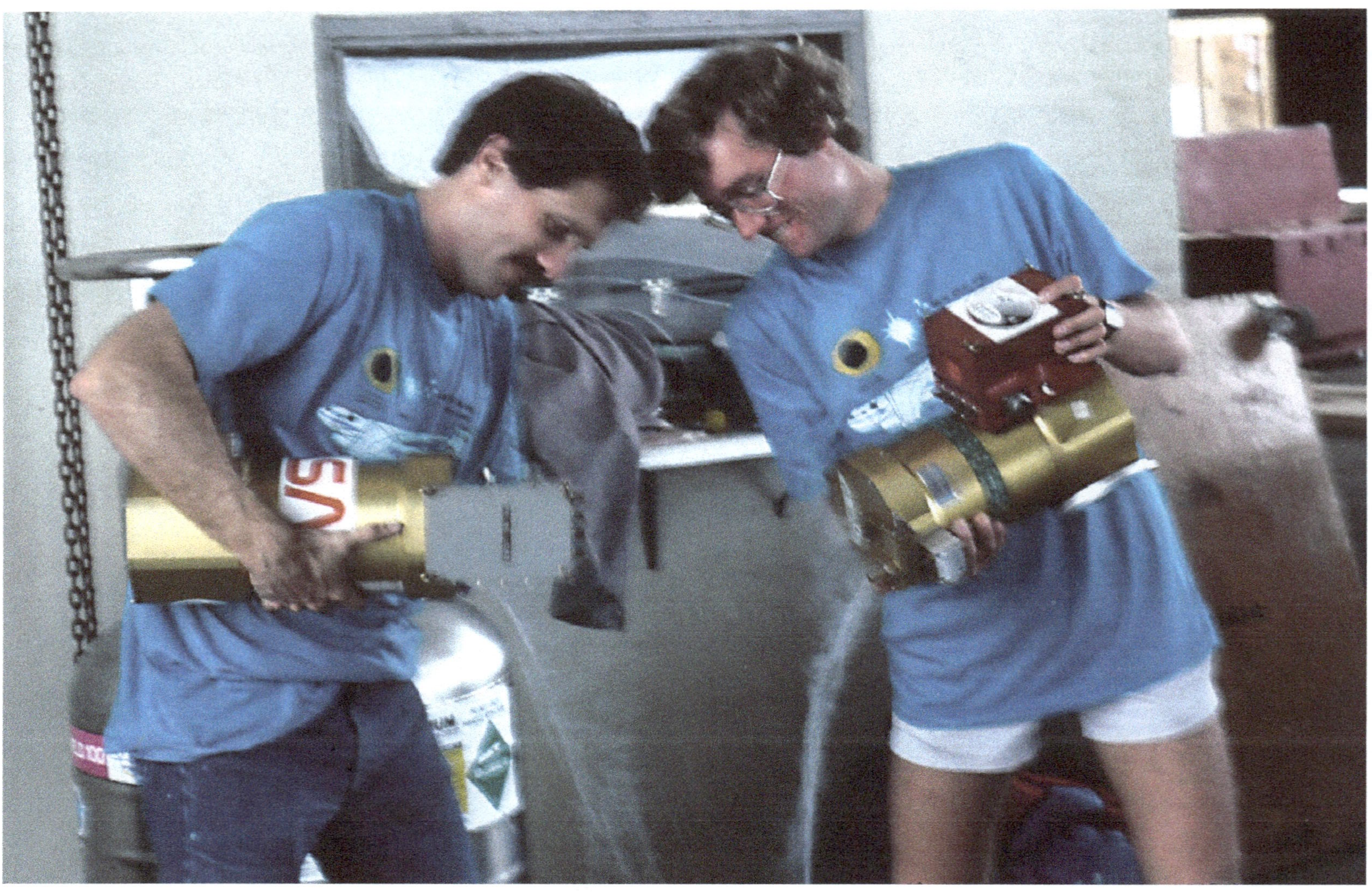

Figure C79. Tom Roellig (NASA Ames) and Stanford graduate student Gregg Kopp, 1988, emptying liquid nitrogen from the detector dewars prior to cooling them with liquid helium. They would use the detectors to make far-infrared observations of a solar eclipse. The expedition was based at Anderson Air Force Base on Guam, where this picture was taken.

1988

Figure C80. Eric Becklin (UCLA), Mike Werner (NASA Ames), Charlie Lindsey (University of Hawaii), 1988. Lindsey developed a full aperture sun filter for the KAO primary mirror. Here he's prepared to observe the solar eclipse before, during, and after totality.

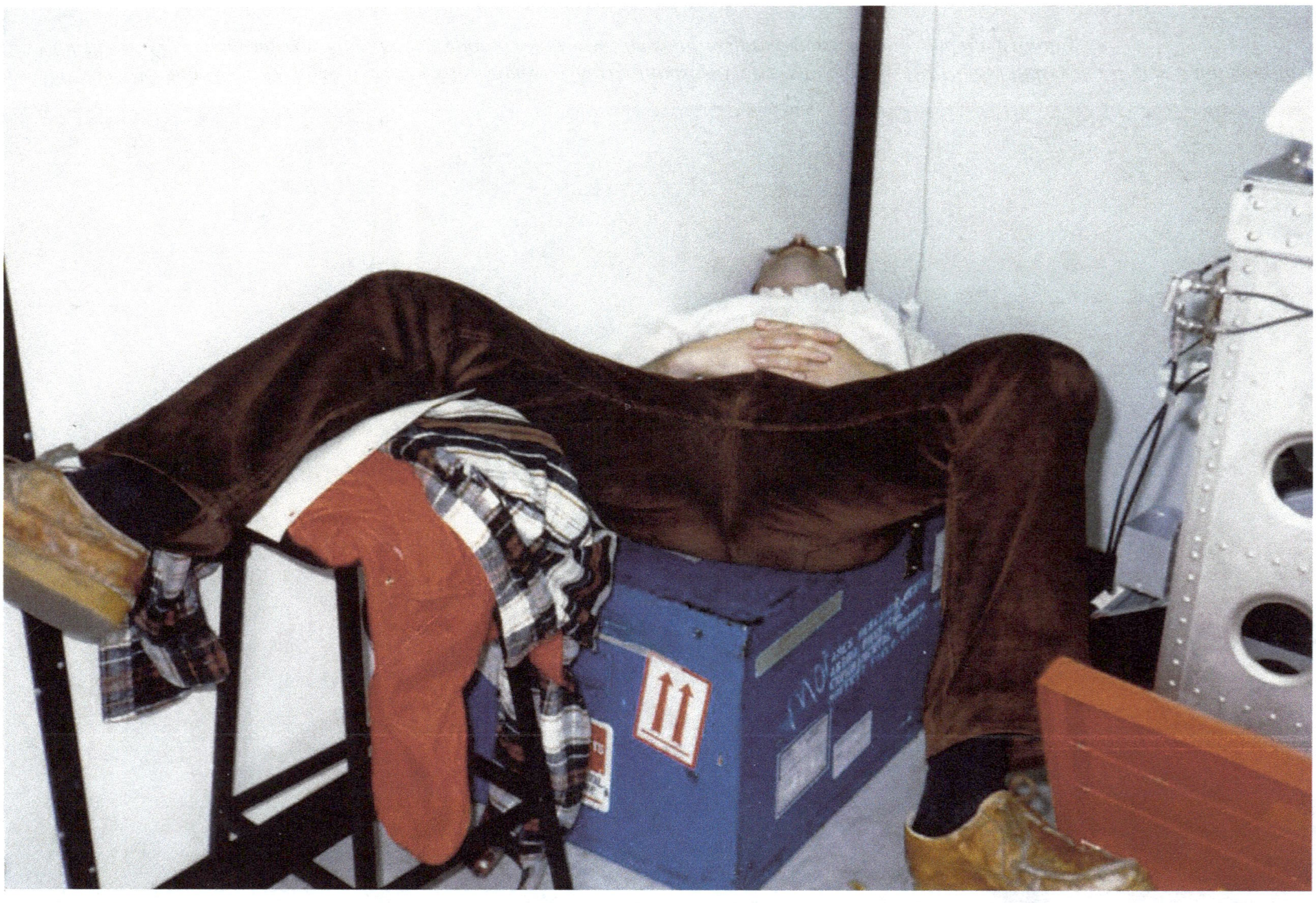

Figure C81. Anonymous scientist napping in the investigators' trailer at the KAO staging site at the Christchurch, New Zealand airport, 1989. Despite proximity of lodgings, schedules sometimes mandated ad hoc rest accommodations.

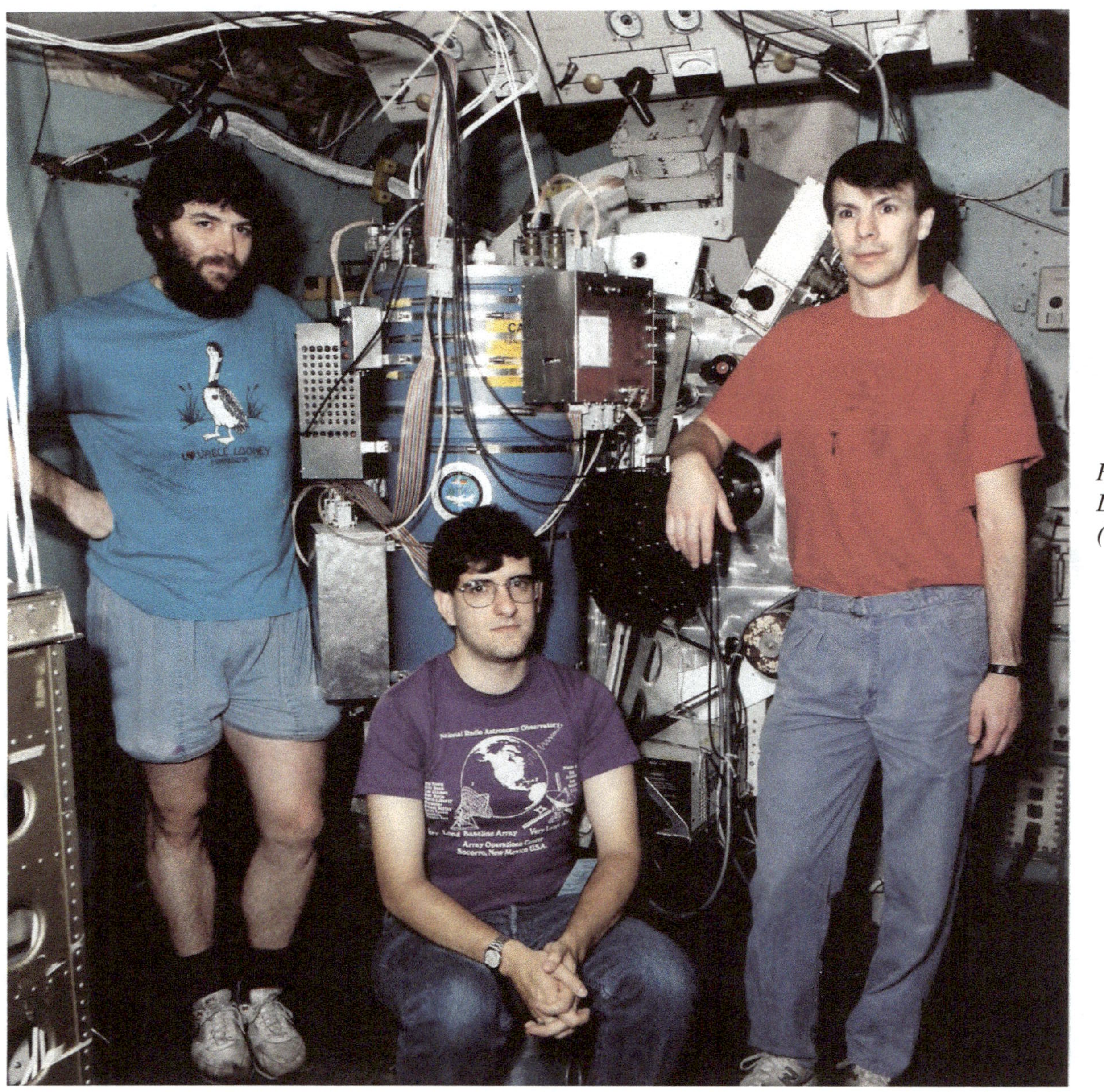

Figure C82. George Gull, student Dave Shupe, Terry Herter (Cornell University), 1990.

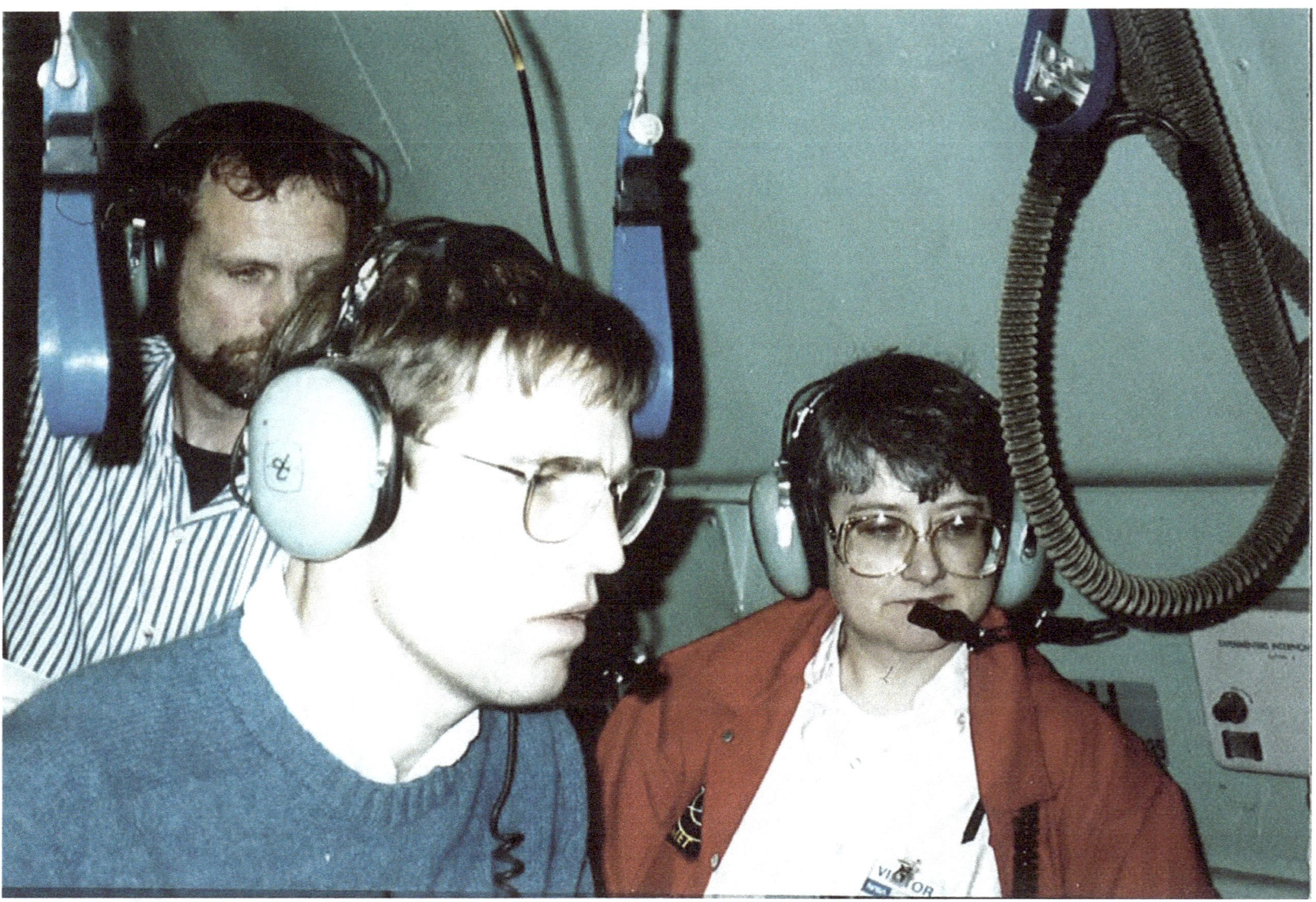

Figure C83. Bob Gehrz (U. Minnesota), Harvey Moseley (Goddard Space Flight Center), and Gehrz's graduate student Carol Stalzer, all focusing on the incoming data stream, 1992.

1992

Figure C84. The KAO on jacks in Ames hangar N-211 for installation of an "aft ramp" on the fuselage behind the telescope cavity, 1992. Ames aerodynamicist Don Buell suggested installing a contoured ramp just aft of the telescope cavity to reattach the air flow over the cavity there. He surmised this configuration would be an improvement over the porous upstream fence previously used to deflect the boundary layer over the cavity. Wind-tunnel tests for SOFIA at Ames verified that this arrangement would be aeromechanically and aeroacousticly superior. The ramp geometry was optimized in these tests.

Figure C85. The KAO on one of two engineering test flights made to evaluate the performance of the "aft ramp" passive flow control (PFC), newly installed behind the telescope cavity, 1992. The new flow-control configuration was highly successful, eliminating all-night vibrations and rumbling of the airframe during observing missions when the door was open in flight.

Figure C86. David Schleuning, Jessie Dotson, Roger Hildebrand, and Darren Dowell (University of Chicago), and Jackie Davidson (NASA Ames post-doctoral associate), 1993. Schleuning and Dotson were graduate students, and Dowell and Davidson former graduate students of Hildebrand, all members of his KAO observing team.

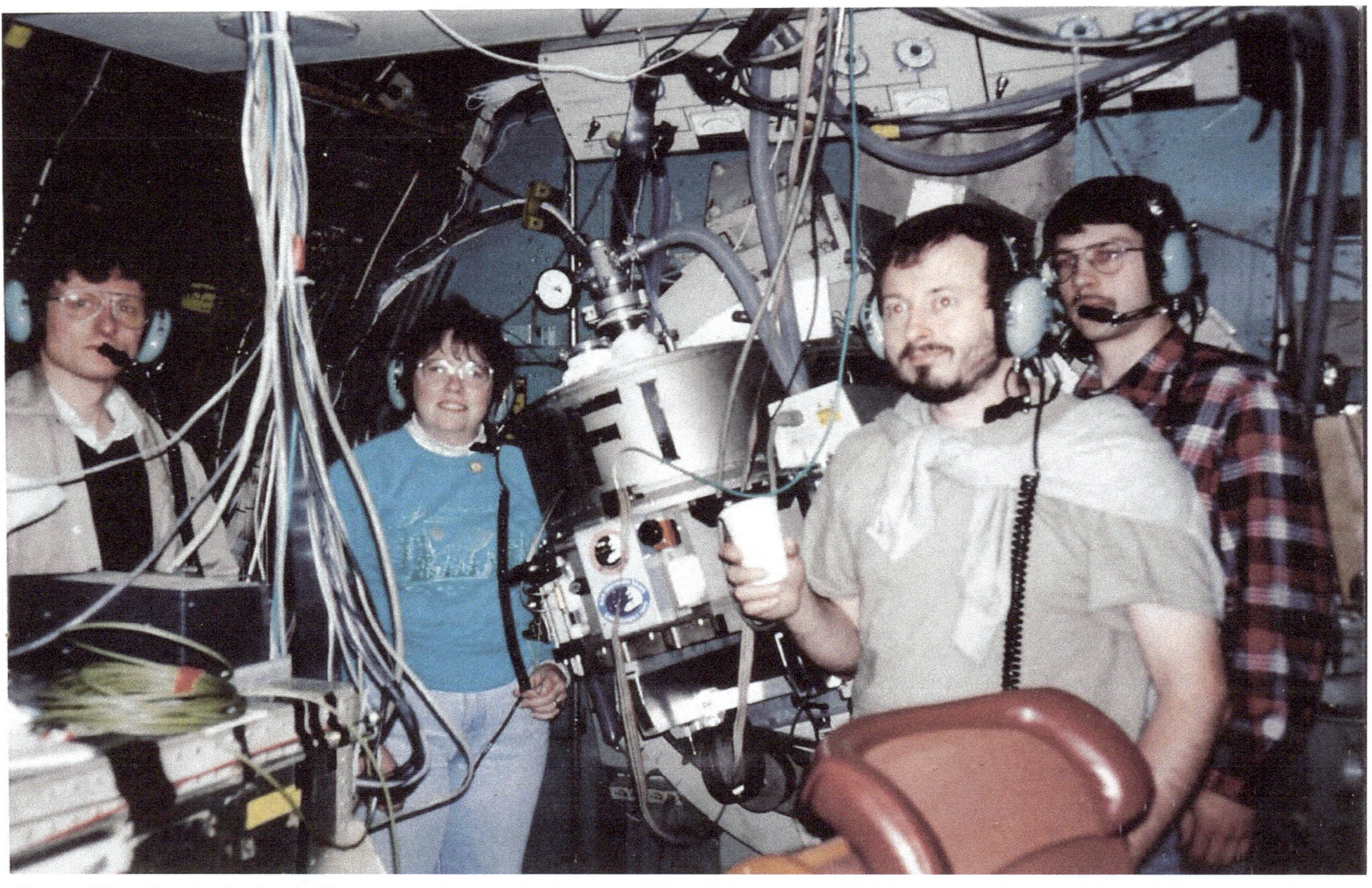

Figure C87. Albrecht Poglitsch (University of California Berkeley, UCB), Edna DeVore (SETI Institute), Norbert Geis, and German graduate student Thomas Nikola (UCB), 1993. The instrument is FIFI, the UCB far-infrared imaging Fabry Perot spectrometer.

Figure C88. Juliette Wiersema, Dora Willoughby, and Sarah Acevedo (NASA Ames) 1994. They helped organize the 1994 Airborne Astronomy Symposium. Willoughby assisted Ames astronomers for years with a variety of research support activities.

Figure C89. Allan and Nancy Meyer, and Bob Cameron (NASA Ames), 1994. Meyer was the chief tracker operator who provided direct support of investigator teams for flight planning over most of the KAO's lifetime. He probably flew more flights than any other staff member. Cameron was the first Medium-Altitude Missions Branch Chief, responsible for the airborne astronomy project at Ames until about 1985 when he left NASA. Cameron was noted for his enthusiasm and witty aphorisms.

Figure C90. Yvonne Pendleton and Ed Erickson (NASA Ames), and Beth Erickson, 1994. Pendleton studied interstellar grain physics with data from the KAO. Ed was the KAO Facility Scientist and a principal investigator on the Learjet and on the KAO throughout its lifetime. Beth supported his work by running a tight domestic ship with their five children.

Figure C91. Hans Mark (University of Texas), and Marianne and Martin Harwit (Smithsonian Air and Space Museum), 1994. While Director of NASA Ames in 1970, Mark established the infrared astronomy program there, and circulated a letter to the astronomical community advocating the development of a large airborne telescope to succeed the KAO. Harwit began infrared observations from the Lear Jet and led a team that observed from the KAO during most of its lifetime.

Figure C92. Larry Caroff (NASA Headquarters), Jackie Davidson (NASA Ames), and Don Osterbrock (University of California at Santa Cruz), at the second decadal Airborne Astronomy Symposium, 1994. Osterbrock was an expert on astrophysical gaseous nebulae, an astronomical historian, and a frequent reviewer of proposals for observations on the KAO.

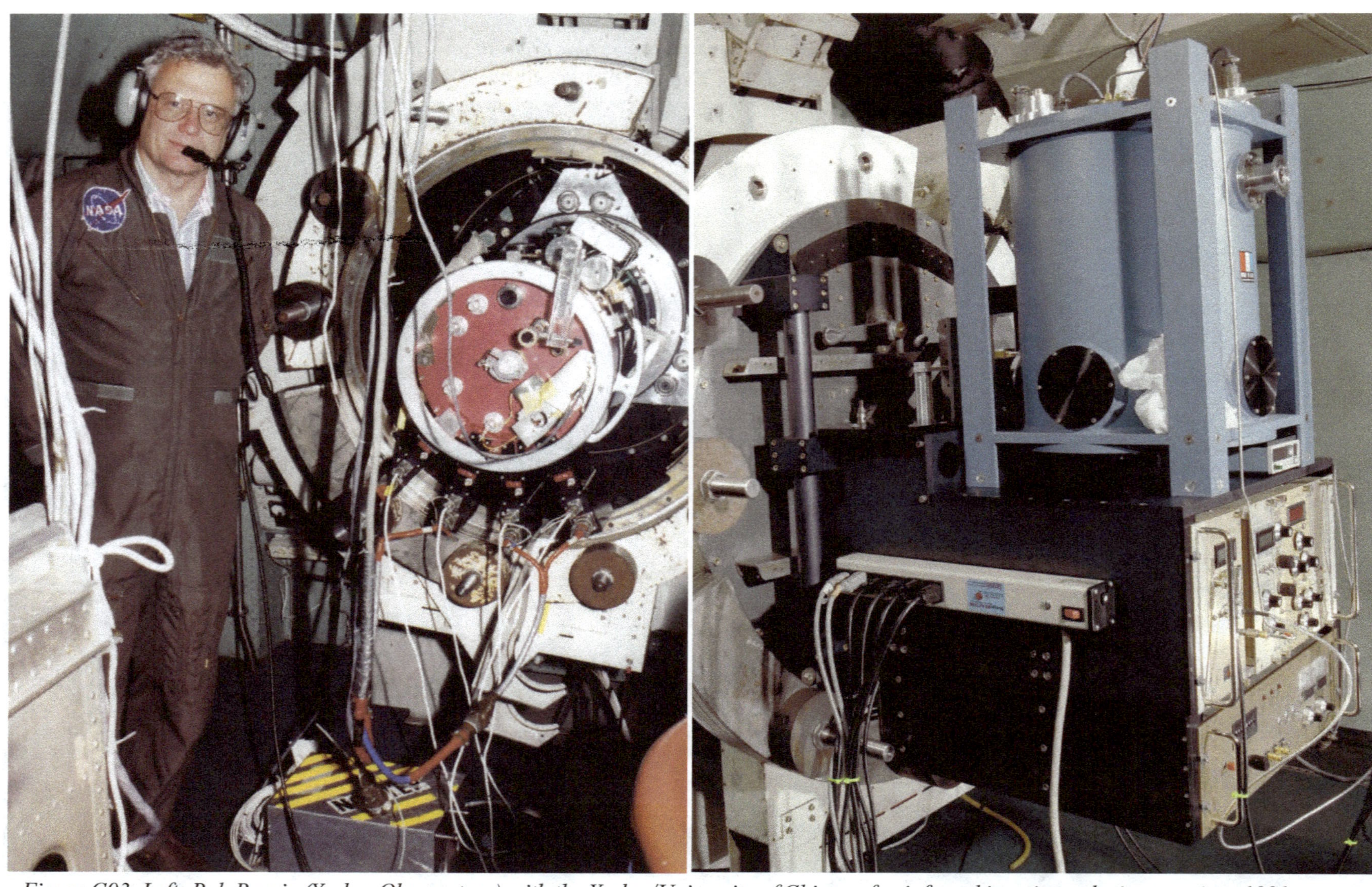

Figure C93. Left: Bob Pernic (Yerkes Observatory) with the Yerkes/University of Chicago far-infrared imaging polarimeter, circa 1991. Right: the submillimeter heterodyne receiver of Jonas Zmuidzinas (Caltech), 1994. These images portray the size and suggest the sophistication to which science instruments had evolved over the operating lifetime of the KAO.

Figure C94. Harry Latvakoski, Leon Peng, Gordon Stacey, George Gull, and Tom Hayward (Cornell University), 1994. Latvakoski and Peng were graduate students. Stacey, as did a few other instrument team leaders in the KAO program, earned his Ph.D. using airborne astronomy observations he made in earlier years. The instrument, KWIC (Kuiper Widefield Infrared Camera), was a mid-infrared imager using a (then) recently developed monolithic charge-coupled device (CCD) array detector.

Figure C95. Don Hunten, Ann Sprague, and Richard Kozlowski (University of Arizona), and Diane Wooden (NASA Ames), 1994. Using the HiFogs mid-infrared spectrometer of Fred Witteborn (NASA Ames), they were observing impacts of Comet Shoemaker-Levy 9 on Jupter, on a KAO flight from Melbourne, Australia.

Figure C96. Back: Tom Barber (Melbourne Airport liaison), Clarence Hoss, Mike Stortz, Ed Mitz, Allan Dunn, Terry Rager, John Brown, Monte Hodges, Danny Rendon, Dan Dorr (U.S. Air Force), and Louie Russo; Front: Gary Morris, Wendy Whiting Dolci, Mark Eshbaugh, Jim Mills, Tom Kolaski, and Chico Rijfkogel, 1994. Except as noted, all were based at NASA Ames. At the time these folks were supporting the flight, ground, and mission operations during the KAO deployment to Melbourne, Australia, to observe the impacts of Comet Shoemaker-Levy on Jupiter. In this intense operation, multiple instruments were used on consecutive nights.

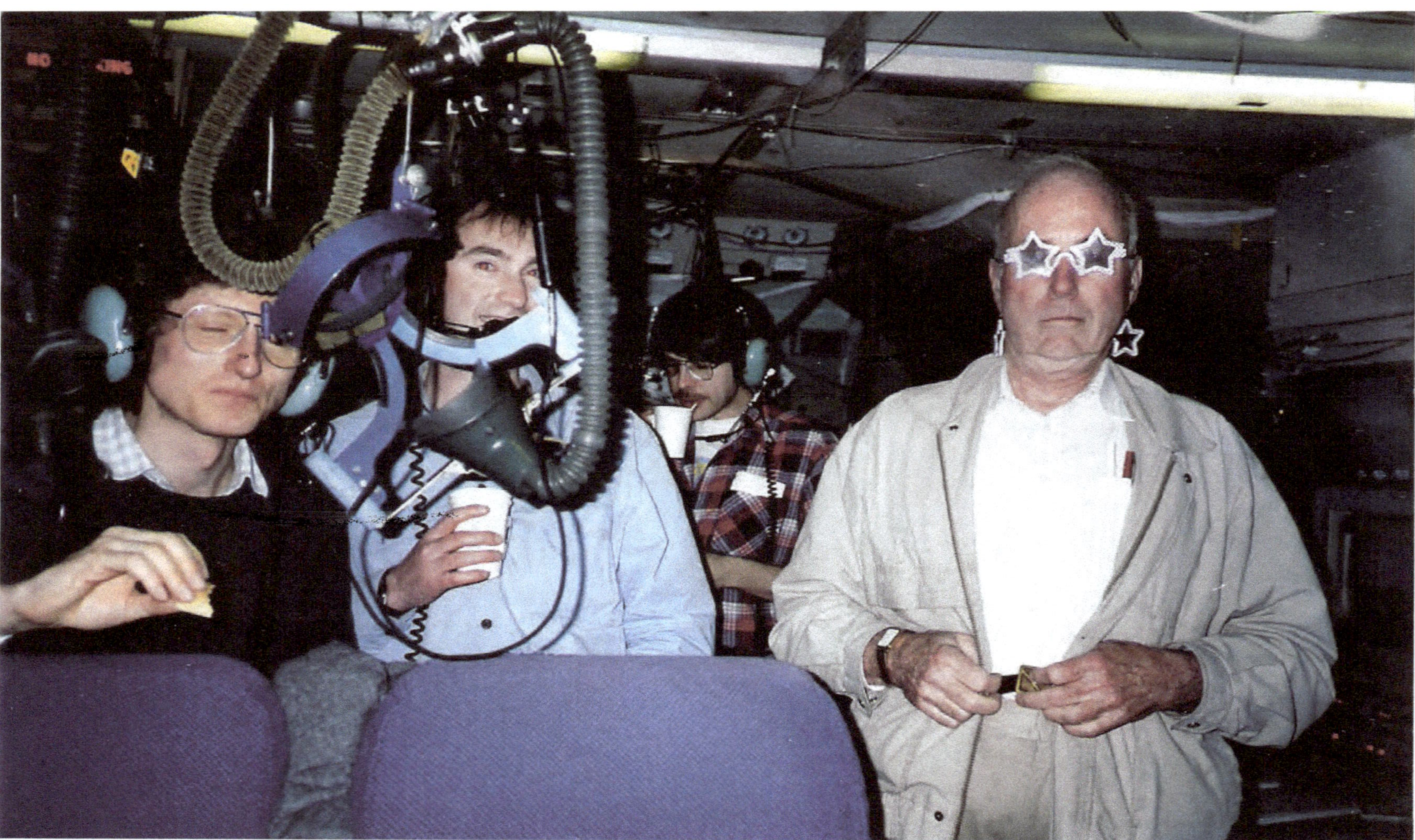

Figure C97. Albrecht Poglitsch (University of California, Berkeley, UCB), post-doctoral associate Ralf Timmermann (Max Planck Institut für Extraterrestrische Physik, Germany), German graduate student Thomas Nikola, and Noble laureate Charles Townes (UCB), 1995. On flights when all was working well, during long integrations on faint objects there were moments for relaxation, lunch, and levity.

Figure C98. Patricia Merha, Edna DeVore (SETI Institute), and Suzanne Maly. Merha and Maly were science teachers from Arizona who were participating in the FOSTER program, which DeVore managed.

Figure C99. Edna DeVore (at left in blue jumpsuit) with FOSTER teachers and assistants, 1995.

1995

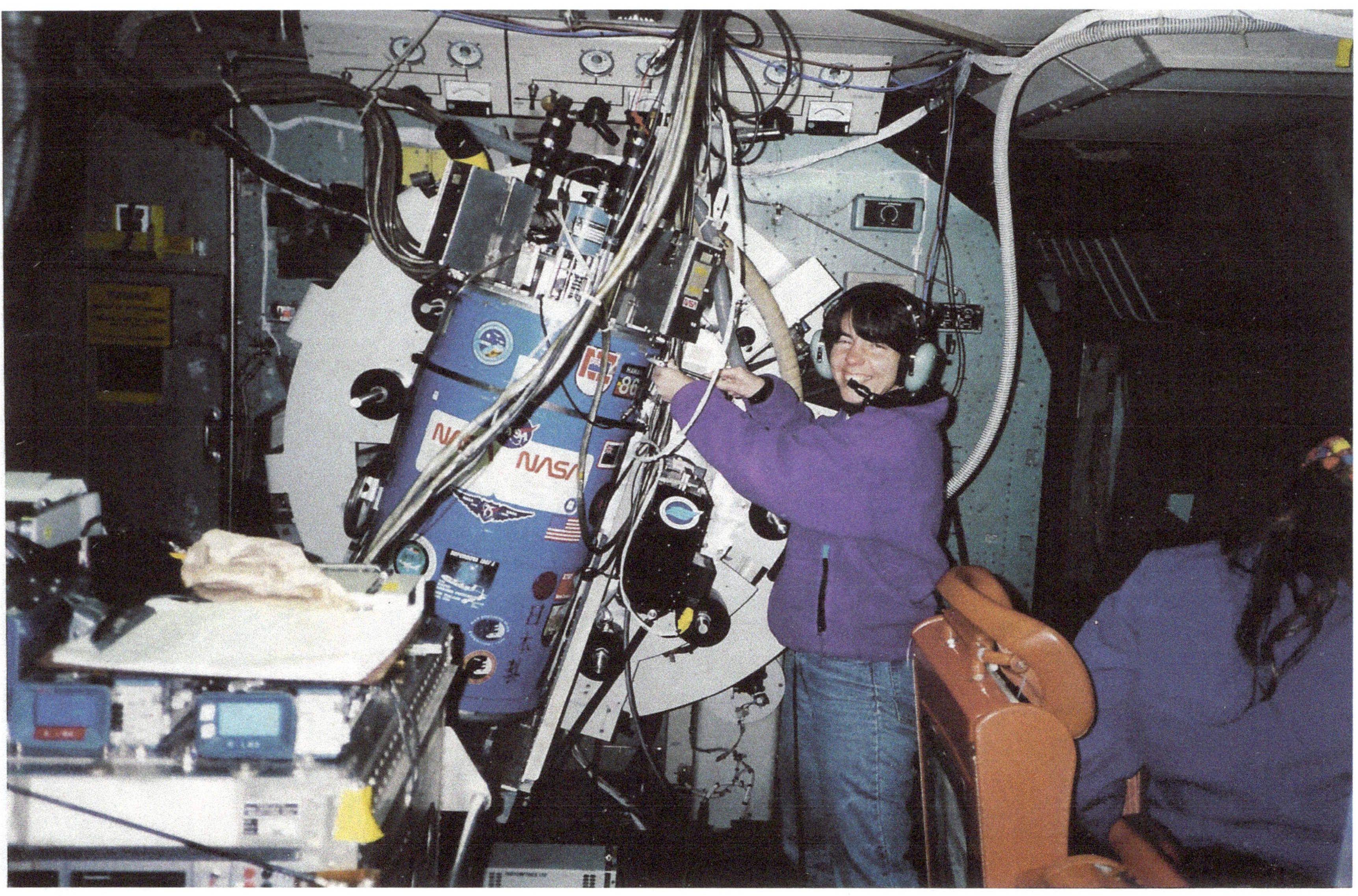

Figure C100. Sue Madden, a postdoctoral associate at NASA Ames, 1995. Here she is seen checking detector biases on the Ames Cryogenic Grating Spectrometer (CGS), just prior to opening the telescope cavity door early in a flight.

Figure C101. Milo Reisner, Jim McClenahan, and Wendy Whiting Dolci (NASA Ames). KAO staff members, Reisner served as a mission director and telescope operator, McClenahan was the de facto project engineer, and Dolci the deputy project manager. This picture was taken at the KAO farewell ceremony, 29 September 1995.

Figure C102. Farewell to the KAO, 29 September 1995. Staff from NASA Ames and many KAO investigators gathered to celebrate their experiences and exploits on this beautiful, unique observatory. Decommissioned at the peak of her productive scientific career to aid the development of SOFIA, she engendered an abiding nostalgia felt by all.

Figure C103. Wendy Whiting Dolci (NASA Ames, seated) and students, 1995. After the last programmed science mission, the KAO was outfitted with communications for an educational program called "Live from the Stratosphere." In October 1995, students and teachers on the ground followed live broadcasts from two flights as data were acquired by a science team. These were the last flights of the KAO. Here, in the KAO on the ground at Ames, students watch as Dolci reviews videos of in-flight action on the observatory.

Figure C104. The KAO (N714NA) and NASA Ames Research Center welcomed their offspring SOFIA (N747NA) at an open house on 14 January 2008. The two observatories are seen here in front of the Flight Operations building at Ames. SOFIA was visiting from its base of operations in Palmdale, California.

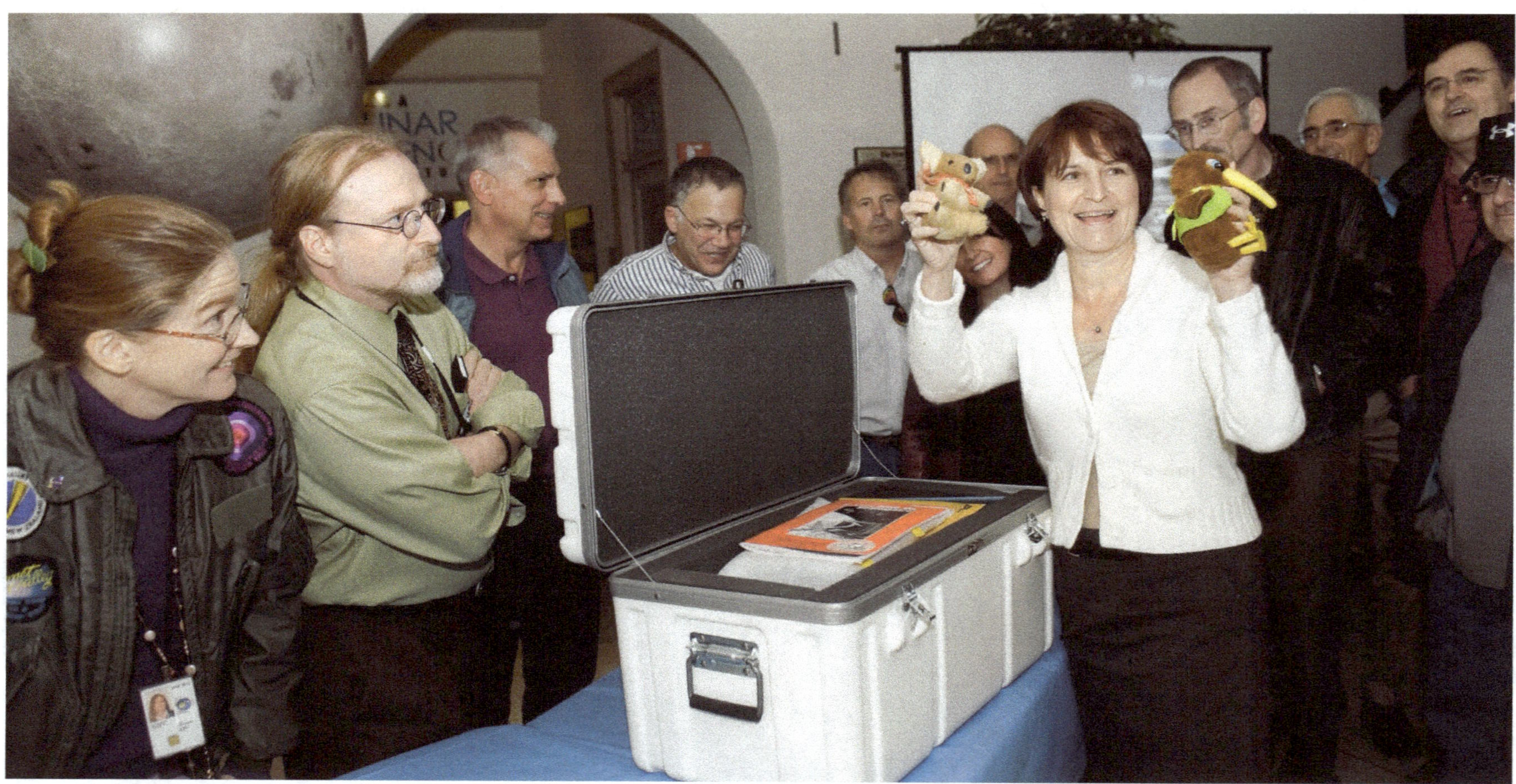

Figure C105. Diane Wooden, Jim Cockrell, Steve Patterson, Allan Meyer, Randy Hobbs, Yvonne Pendleton, Lou Allamandola, Wendy Whiting Dolci, Mike Haas, Ed Erickson, Steve Culp, and Louie Russo, NASA Ames 10 November 2010. The occasion was the opening of the KAO Time Capsule that was sealed on the last flight of the KAO in October 1995. Wendy is displaying mascots that flew on the KAO. Diane, Yvonne, Lou, Mike, and Ed were KAO investigators; Jim, Allan, Wendy, and Steve were Observatory Staff; Steve Patterson, Randy, and Louie were KAO ground crewmen.

Figure C106. Reunion of some former participants in the KAO program, with family members and friends, held at NASA Ames on 13 August 2011. All have aged, most at least more gracefully than the old lady, which has sat idly on the ramp since 1995.

RFERENCES

1. Haas, M.R.; Davidson, J.A.; and Erickson, E.F. (eds): Airborne Astronomy Symposium on the Galactic Ecosystem: From Gas to Stars to Dust. ASP Conference Series, vol. 73, 1995.
2. Larson, H.P.: The NASA Airborne Astronomy Program: A Perspective on Its Contributions to Science, Technology, and Education. PASP, vol. 104, pp. 146–153, 1992.
3. Thronson, H.A. and Erickson, E.F. (eds): Airborne Astronomy Symposium, NASA Conference Publication 2353, 1984.
4. Erickson, E.F.: The Sofia Program. Proceedings of The Dusty and Molecular Universe, A Prelude to Herschel and Alma. European Space Agency SP-577, p. 69, Jan. 2005.
5. Erickson, E.F. and Davidson, J.A.: SOFIA: The Next Generation Airborne Observatory. ASP Conference Series, vol. 73, p. 707, 1995.
6. Dolci, W.W.: Milestones in Airborne Astronomy: From the 1920's to the Present. AIAA paper no. 975609, 1997.
7. Kuiper, G.P. and Forbes, F.F.: High Altitude Spectra From NASA CV 990 Jet I: Venus, 1-2.5 Microns, Resolution 20 cm^{-1}. University of Arizona Communications of the Lunar and Planetary Laboratory, no. 95, 1967.
8. Low, F.J.: Airborne Infrared Astronomy: The Early Days. Airborne Astronomy Symposium, Harley A. Thronson and Edwin F. Erickson, eds., NASA Conference Publication 2353, pp. 1–8, 1984.
9. Erickson, E.F.; Goorvitch, D.; Dix, M.J.; and Hitchman, J.J.: Learjet Telescope System. NASA TM X-62, p. 389, 1974.
10. Mastenbrook, H.J.: Water Vapor Distribution in the Stratosphere and High Troposphere. Journal of Atmospheric Science, vol. 25, p. 299, 1968.
11. NOAA, NASA, USAF, and National Weather Bureau. U.S. Standard Atmosphere, GPO, Washington, D.C., 1976.
12. Lord, S.D.: Modeling Earth Atmospheric Transmittance of Near-Infrared and Far-Infrared Radiation. NASA Technical Memorandum 103957, 1992.

13 Erickson, E.F.: Effects of Telluric Water Vapor on Airborne Infrared Observations. PASP, vol. 110, pp. 1098–1105, 1998.

14. Cameron, R.M.; Bader, M.; and Mobley, R.E.: Design and Operation of the NASA 91.5-cm Airborne Telescope. Applied Optics, vol. 10, p. 2011, 1971.
15. Erickson, E.F. and Matthews, S.: WINDOW: A computer Program for Planning Astronomical Observations. NASA TM X-73156, 1976. Later versions were entitled "WINDO"; see ADAMS (Airborne Data and Management System) User's Handbook 1991, KAO User's Handbook 1994.
16. Fiekowski, P.: [KNAV] Flight Planning Tutorial. ADAMS User's Handbook 1984; also ADAMS User's Handbook 1991, KAO User's Handbook 1994.

17. Haas, M.R. and Phister, L.: A High-Altitude Site Survey for SOFIA. PASP, vol. 110, pp. 339–364, 1998.

18. Erickson, E.F.; Baltz, J.A.; and Morey, B.C.: Information Displayed on the KAO Video Distribution System: Implications for SOFIA. SOFIA Technical Report 1991.

19. Elliot, J.L.; Dunham, E.T.; and Mink, D.: The Rings of Uranus. Nature, vol. 267, p. 328, 1977.

20. Haas, M.R.; Hollenbach, D.J.; and Erickson, E.F.: Detection of [Si II] (34.8 Micron) Emission in Orion-KL: A Measurement of the Silicon Abundance in Dense Interstellar Gas. Astrophysical Journal, vol. 301, p. L57, 1986.

21. http://en.wikipedia.org/wiki/DO-178B, Software Considerations in Airborne Systems and Equipment Certification. Accessed Jan. 14, 2014.

22. Keller, J. and Williams, S.: Bringing Students Together: the Impact of the FOSTER Program for Teachers. Airborne Astronomy Symposium on the Galactic Ecosystem: From Gas to Stars to Dust. Michael R. Haas, Jacqueline A. Davidson, and Edwin F. Erickson (eds.), ASP Conference Series, vol. 73, p. 635, 1995.

23. Koch, D. and Cox, S.: SOFIA: Derivation of Operations Costs. NASA Internal Report, 1991.

24. Fusco, P.: private communication, 2008.

25. Dunham, E.W.; Young, E.F.; Baron, R.L.: Elliot, J.L.; Griffith, M.; and Wastts, A.W.: Investigation of the Images Formed by the Telescope in the Kuiper Airborne Observatory. Final Report for June 1986 Flight Series Grant Nag2-257, 1987.

26. Colgan, S.W.J.; Erickson, E.F.; Haynes, F.B.; and Rank, D.M.: A CCD Offset Guider for the KAO. Airborne Astronomy Symposium on the Galactic Ecosystem: From Gas to Stars to Dust. Michael R. Haas, Jacqueline A. Davidson, and Edwin F. Erickson (eds.), ASP Conference Series, vol. 73, p. 567, 1995

27. http://www.bls.gov/data/inflation_calculator.htm. Accessed Jan. 14, 2014.

28. Hobbs, R.: private communication, 2013.

29. Shen, L.: private communication, 2013.

30. Davidson, G.: Potential Measures of the Scientific Productivity of Astrophysics Missions. NASA Headquarters Astrophysics Division Report, 1993.

31. Connors, C.: Lessons from the KAO. NASA SP-781, 2000.

32. Rentch, B. and Zaitzeff, E.: Kuiper Airborne Observatory (KAO) Personnel Interview Chronicle. SPO SER BWR 001 Rev B, Nov. 21, 1996.

33. Stacey, G.J.: private communication, 1995. Over 100 astronomers, the majority of whom were KAO investigators, recommended co-located personnel and facilities to operate SOFIA at Ames.

34. NASA Office of the Inspector General: Final Report on Audit of Aircraft Consolidation at the Dryden Flight Research Center (DFRC). HA-96-007, 1996.

35. Stein, W.A. and Woolf, N.J.: Astronomical Infrared Telescopes. Applied Optics, vol. 10, p. 655, 1971.

36. http://en.wikipedia.org/wiki/Spitzer_Space_Telescope. Accessed Jan. 14, 2014.

37. Astronomy Survey Committee: Astronomy and Astrophysics for the 1970s. National Academy Press, Washington D.C., 1972.

38. Astronomy Survey Committee: Astronomy and Astrophysics for the 1980s. National Academy Press, Washington D.C., 1982.

39. Sutton, G. W. and Pond, J. E.: Predictions of the SOFIA Telescope Seeing in Flight. Optical Engineering, vol. 37, p. 2872, 1998.

40. Astronomy Survey Committee: The Decade of Discovery in Astronomy and Astrophysics. National Academy Press, Washington D.C., 1991

41. Erickson, E.F., and Meyer, A.W.: SOFIA Science Objectives, Rationale, and Related Requirements. SOFIA Project Document SOF 1009, 1997.

42. Erickson, E.F. et al.: Co-authored by 50 infrared astronomers, this paper describes the need and potential for "Training of Instrumentalists and Development of New Technologies on SOFIA." http://arxiv.org/ftp/arxiv/papers/0903/0903.4240.pdf, 2009. Accessed Jan. 26, 2014.

CITED PARTICIPANTS

Individuals referred to and / or pictured are listed here with a reference to their locations in this document. The abbreviations are: A = acknowledgements; ded = dedication; B, C, or D = appendices; and numbers = section of the paper.

Munar, Lori	A
Mundy, Lee	B
Muser, Dieter	5.2
Myers, Mike	B
Natta, Antonella	C
Neugebauer, Gary	C
Ney, Ed	B
Nikola, Thomas	C
Noll, Keith	B, C
Novak, Giles	B, C
Oishi, Don	C
Olson, Don	C
Omont, Allain	B
Osterbrock, Don	C
Panteleo, Jim	C
Pants, Fred	C
Parmer, Parvinder	C
Patterson, Jim	C
Pellerin, Charlie	5.2
Pendleton, Yvonne	B, C
Peng, Leon	C
Pernic, Bob	C
Petuchowski, Sam	B
Pfister, Lenny	4.5
Phillips, Tom	4.1, B, C
Pickett, Herb	C
Pipher, Judy	B, C
Poglitsch, Albrecht	B, C
Pollock, Dave	C
Poppen, Richard	C
Puetter, Rick	C
Ragasa, Al	C
Rager, Terry	C
Rank, Dave	B
Reichenbach, Mike	C
Reisner, Milo	4.7, C
Rendon, Danny	C
Retig, Dave	C
Rhoades, Chris	C
Rice, Carl	C
Rickard, Lee	C
Rickets, Bill	C
Rijfkogel, Chico	C
Ritter, Tex	C
Roellig, Tom	A, B, C
Roeser, Hans-Peter	4.8, 5.2, B, C
Rose, Bill	5.2, C
Rubin, Bob	B
Rudolph, Alex	4.6, B
Russell, Ray	4.8, B, C
Russo, Louie	C
Sandford, Scott	C
Saykally, Richard	B
Schaefer, Frank	C
Schaeffer, Mark	C
Schleuning, David	C
Scoville, Nick	B
Sellgren, Chris	4.1, C
Shaya, Ed	C
Shen, Leon	A
Shupe, Dave	C
Shure, Mark	C
Simon, Mike	B
Simpson, Jan	C
Skinner, Chris	B
Skrutskie, Mike	B, C
Smith, Howard	B, C
Smith, Jim	C
Smoot, George	2.3
Soifer, Tom	B, C
Sprague, Ann	B, C
Stacey, Gordon;	4.8, B, C
Stalzer, Carol	C
Stoeffler, Terry	C
Storey, John	4.1, B, C
Stortz, Mike	C
Strecker, Don	C
Strelnitski,Vladimir	B
Strom, Steve	B
Stutzki, Jurgen	B
Sutton, George	5.2
Swanson, Paul	C

ABOUT THE AUTHORS

Ed Erickson earned a Ph.D. in physics with minor in mathematics at Stanford University. His dissertation topic was scattering of high energy electrons from deuterium: experimental and theoretical research designed to explore the structure of the neutron and the interaction of neutrons and protons at small distances. After 2 years on active duty with the U.S. Army, part of which was spent in the Instrumentation Branch at NASA Ames, Erickson did particle physics experiments in Europe for 4 years before returning to California with his family in late 1970. He joined the Astrophysics Branch at Ames in early 1971, developing instrumentation for, and making infrared observations from, the NASA Learjet and subsequently from the Kuiper Airborne Observatory, which began operations in 1974.

Most of his research with colleagues at Ames involved far-infrared spectroscopy from the KAO, for which he was NASA's Facility Scientist. Some of the findings of this research included evidence for sulfuric acid aerosols comprising the clouds of Venus, thermal structure of star-forming molecular clouds, existence of hot stars in the vicinity of the Galactic Center, production of heavy elements from Supernova 1987A, and internal heat sources in Jupiter and Saturn. He was a member of the Near Infrared and Multi-Object Spectrometer (NICMOS) instrument team for the Hubble Space Telescope, which he and collaborators used to study shocked gaseous structures driven by luminous objects in the Orion molecular cloud, to identify young stellar objects in other regions of massive star formation, and to probe effects of magnetic fields in outflow regions around young stars.

Erickson devoted a significant part of his later career to promoting, providing technical analyses for, and specifying requirements for SOFIA; he organized the SOFIA Science Working Group and became NASA's original Project Scientist. He retired from NASA in 2007 but still consults on SOFIA-related issues. Married with five children, he enjoys visiting, vacationing, and projects with them; cycling; ballroom dancing; coaching youth soccer; talking to school groups about astrophysics; and completing long overdue improvements to his home in Sunnyvale, California.

Allan Meyer obtained a Master's Degree in Astronomy and Astrophysics at the University of California, Santa Cruz. Following graduation in 1974, he took a position as a data analyst in the Space Science Division at NASA Ames, assisting with research on meteorite composition, impact crater simulations and infrared spectroscopy. From 1977 until 1995 he was a staff member of the Kuiper Airborne Observatory. There he developed custom techniques and trained others to enable accurate, reliable telescope performance—target acquisition and fine pointing—under the time-critical flight conditions of airborne astronomy. In this role he flew more flights than any other KAO crew member, including missions to observe Halley's comet, SN 1987A, far-infrared mapping of the Magellanic Clouds, the discovery of the rings of Uranus, and the discovery of Pluto's atmosphere.

As a staff scientist for the KAO, he assisted visiting astronomy teams in planning efficient observing missions. His catalog of objects observed from the KAO and those from spacecraft (IRAS, HST) data archives were used frequently in selecting astronomical targets for otherwise "dead" flight legs, or when in-flight changes to the flight plan were required. In addition to astronomy flights, Meyer participated in Space Shuttle chase missions and classified DoD/DoE missions on the KAO. Working with the SOFIA Project Scientist from 1996 until 2002, he participated in development of SOFIA science and telescope requirements, consulted extensively with the German telescope developers, and helped plan SOFIA science operations. As an ex-officio member of the SOFIA Science Working Group, he assisted scientists advising the NASA Project Office on a variety of technical issues.

In 2002 he joined the prime contractor (USRA) team for development and operation of SOFIA, where he continues as an Associate Scientist. His SOFIA work relates to obtaining maximum performance from the observatory, including telescope testing and improvements, subsystems, software, flight planning, and advising and training others. Meyer has authored or co-authored research publications, including infrared spectroscopy of Saturn from Cassini, infrared spectroscopy of OH masers, and infrared scattering properties of materials. In 2013 he was awarded a doctoral degree by the University of Stuttgart, Germany, in recognition of his career of contributions to the success of airborne astronomy. Married with three daughters, he enjoys holidays with his family and sailing his 22-foot sloop with them and others. His home is in Campbell, California.

POSTSCRIPT

Those who participated in the airborne astronomy program at NASA Ames recall it with fond memories of challenging work, camaraderie, remarkable moments, and unique scientific achievements. We hope that 20 or more years from now, those who have participated in the development and operation of SOFIA will savor similar memories.

www.ingramcontent.com/pod-product-compliance
Lightning Source LLC
Chambersburg PA
CBHW080542220326
41599CB00032B/6332
* 9 7 8 1 7 8 2 6 6 7 2 3 0 *